Innovative Shear Design

Innovative Shear Design

Hrista Stamenkovic

CRC Press
Taylor & Francis Group
Boca Raton London New York

CRC Press is an imprint of the
Taylor & Francis Group, an **informa** business

CRC Press
Taylor & Francis Group
6000 Broken Sound Parkway NW, Suite 300
Boca Raton, FL 33487-2742

First issued in paperback 2019

© 2002 by Taylor & Francis Group, LLC
CRC Press is an imprint of Taylor & Francis Group, an Informa business

No claim to original U.S. Government works

ISBN-13: 978-0-415-25836-4 (hbk)
ISBN-13: 978-0-367-86362-3 (pbk)

Visit the Taylor & Francis Web site at
http://www.taylorandfrancis.com

and the CRC Press Web site at
http://www.crcpress.com

This study is dedicated to my mother
Kata Bozilovic-Stamenkovic
whose breath of life is woven into this work
Her *Hrista*

An appreciation

Special thanks to Dr. V. Ramakrishnan, Regents Distinguished Professor of Civil Engineering, South Dakota School of Mines and Technology, for his continued help and encouragement given to me in writing and publishing this book and for his constructive and valuable suggestions for improving the manuscript. He is the first person who accepted and approved the new theory enunciated in this book.

Contents

Preface

This study provides evidence that the internal active and internal resisting forces of compression, tension, horizontal and vertical shear are created only in flexurally bent members, as a result of the action of flexural bending forces. Further, it is a proven fact that the combined action of internal active forces results in diagonal cracks and diagonal failure of flexurally bent members. Because the internal active forces of compression and tension are equal in magnitude to active vertical shear forces, at a distance from the support of about the height of the beam, it is evident that this cross section is the most critical. The first diagonal cracking will occur precisely at that location. At the same time, it is proved that the assumption, by Ritter and Morsch (in 1898 and 1902, respectively), of the existence of pure shear leading to diagonal tension at the bent member, is in essence, unprovable and does not exist.

It should be emphasized separately that the existence of internal active and internal resisting forces, in a flexurally bent member, is not covered by either Newton's first or third law. Consequently, it is a new law of physics itself, as well as for engineering mechanics, in general.

We have also included the details of the graphical and mathematical determination of such forces in any bent member as well as the corresponding stresses in such a member. We also show how to control such stresses to prevent cracking and failure of a given member for a given load. We explain, very clearly, that the new law of physics simultaneously coexists with Newton's third law in the same bent member, and that they supplement each other.

Observing such forces, as they are, it is very easy to visualize that the classical theory of calculation of a shear wall, as a vertical cantilever, is untenable. This is because a real cantilever is part of a beam on two supports (rotated by 180°), where supports (reactions) are the concentrated loads at its end. The fixed cross section of such a cantilever is located where shear forces are zero, for a beam on two supports. This fact means that diagonals for a real cantilever are located, for a beam on two supports, where shear forces are zero (Figure 2.10). This also means that diagonals for a real cantilever are not exposed simultaneously to elongation and shortening during loading conditions. On the contrary, one diagonal is exposed to elongation (tension) and the other to shortening (compression); therefore, the shear wall is exposed to pure shear, while a real cantilever is exposed to flexural bending. These two facts lead us to a revolutionary new concept of designing a shear wall. Instead of vertical and horizontal reinforcement being used (according to the truss analogy theory), diagonal reinforcement is applied for these walls. By so doing, forces act longitudinally through the bars, achieving the highest level of reinforcement exploitation. At the same time, the detailing of reinforcement is so simplified that it is needed for only one bar in several layers. Consequently, bad reinforcement detailing is eliminated completely.

Regardless of the slenderness of the shear wall, any such cantilevered wall could be designed as a shear wall triangularly reinforced and not as a cantilever beam with the neutral axis in its body. Such rationale is supported by structural calculations, corresponding figures, explanation, and laboratory testing.

The contribution of the action of internal active compression, tensile and shear forces to the fatigue failure caused by vibrations of the given element was not known. We emphasize that the points of inflection, due to the vibration of the beam, are the nuclei of the concentration of the residual stresses, induced by the transition of the compression zone into the tensile zone, and the contribution of the action of the vertical shear forces, acting at the point of inflection.

In addition, by applying this new law of physics, it becomes possible for the first time in engineering science to explain why prestressed concrete could be designed to, literally, be immune to any diagonal cracking and failure, which is of utmost importance to the prestressed concrete industry in the world.

In Section 5.10 of Chapter 5, the mysterious failures and unpredictable cracks occurring in thousands of steel-framed structures, during the 1994 earthquake in Northridge, California, are explained quite clearly. This was, in fact, caused by the action of internal active compression, tensile and shear forces, and not by the action of diagonal tension which, as such, does not exist.

It is clearly proved that the concept of truss creation in a flexural bending member is simply an illusion, as long as compression and tensile zones coexist, and is irrelevant to a cracked or uncracked member. This means that the entire basis of structural calculations, relying on the Truss Analogy Theory, is untenable. In addition, it is proved that the diagram of shear forces (stresses) of a simply supported beam, could never be identical to the shear diagram of a fixed end beam. For that reason, a new shear diagram is incorporated for a fixed end beam, where forces in the diagram follow the real stressed condition of a flexurally bent member.

By transferring the load of the stringer to the top of the beam, an inverted T beam will be much safer to design concrete bridges in the future. This concept is presented as a separate paragraph in this book.

The author owes special thanks to Professor Gordon Batson of Clarkson University, Potsdam, New York, who reviewed this entire book and gave many excellent suggestions for improvement of this manuscript. The author also wishes to thank Professor Vitelmo Bertero of the University of California, Berkeley, for his review of the chapter on shear wall and for giving many suggestions for the improvement of the chapter. In addition, the former Deputy Director of the Public Works Department in the City of Riverside, California, Mr. William D. Gardner, P.E., made numerous suggestions for improvement of the language and deserves special thanks. And last, but not the least, Miss Patricia Ames, an engineer in the Public Works Department of the City of Riverside, California, has earned enormous thanks and appreciation for her devoted typing and editing of the language in this manuscript. Without her help the publication of this book would not have been possible. Also, special credit belongs to Mr. Donald P. Young, P.E., of the same Public Works Department for typing the final version of this book.

September 1998   Hrista Stamenkovic
                                                                          Riverside, CA

Foreword

It is with great pleasure and honour that I was asked to read "Innovative Shear Design" by Dipl Ing Hrista Stamenkovic. I have read a number of treatises on diagonal failure in reinforced concrete over the 40 years of my academic career. There is no shadow of a doubt that Mr. Stamenkovic's work on this topic is unique and original.

Mr. Stamenkovic presents a new law of physics for a bent member. The new law introduces the concept of "Internal Active Forces" and "Internal Resisting Forces", which are essential for the equilibrium of a free body diagram. The concept is an exciting new development and helps in understanding the behaviour of a bent member. It also leads to rational solutions of many problems associated with a bent member.

Mr. Stamenkovic's shear design method based on his new law of physics is fresh and innovative. In his work, he identifies two different critical cross-sections in a flexurally bent member, namely (a) a critical flexural cross-section, where the bending moment is the largest and failure is expected to occur due to tension and compression; and (b) a critical shear cross-section where internal active compression, tensile and shear forces become equal in magnitude and failure is expected to occur under the action of resultant punch shear force.

The method leads to efficient reinforcement against diagonal failure in bent members. It is easy to follow and ready to be put to use. The code writing authorities should give serious consideration to adopting the method into the building codes.

Mr. Stamenkovic sheds light on the real cause of fatigue failure. He demonstrates by means of a simple experiment that the point of inflexion (PI) in a bent member is the most vulnerable location for fatigue failure to occur. The failure is due to three different factors, namely, punch shear force (or vertical shear force) at the PI, plastic deformation in the cross-section, and temperature increase which decreases the yield strength of the material. His arguments are very lucid and valid.

Using his new law of physics, Mr. Stamenkovic produces very sensible shear force diagrams for continuous or fixed-end beams. A fixed-end beam is considered to be made of a simple beam and two cantilevers. The PI serves as a statical support for the simple beam. The shear force diagram of the simple beam with the applied loads is then constructed. The shear force diagram of the cantilever with applied loads and the end shear force equal in magnitude but opposite in direction to shear force at the end of the simple beam is also drawn. These two diagrams are finally combined to produce the shear force diagram for the fixed-end beam. The net result is an exciting new shear force diagram, which represents the real orientation and magnitude of shear forces at any cross-section. The new diagram shows a change in the direction of shear forces at a PI and therefore clearly demonstrates the extremely sensitive nature of that cross-section.

The ingenious application of the author's law of physics is once again revealed in the design of shear walls or shear membranes. His triangular reinforcement method is rational and most efficient. The superiority of the method is clearly demonstrated in the wall test he conducted. It is no surprise that the wall designed according to his method resisted nearly double the load carried by a similar wall with conventional horizontal and vertical reinforcement. This triangular reinforcement method not only provides resistance to loads but also provides the rigidity that is essential in walls or membranes subjected to seismic effects. The comprehensive design example and other details given in the book will be very useful for designers.

In all, I thoroughly enjoyed the work of Dipl Ing Hrista Stamenkovic and am enlightened by it. I fully appreciate, understand, endorse and recommend this book to professional engineers, academics and students, with the advice that the reader should approach it with an open mind.

I sincerely wish that the book finds complete acceptance in the engineering community and that Mr. Stamenkovic wins wide acclaim for his ingenious work.

Dr B Vijaya Rangan, FACI
Dean of Engineering
Curtin University of Technology
Perth, Australia

Chapter 1

Introduction

Professor Gustav Magnel (Belgium), in his lecture in the United States on 5 February 1954, presented his opinion, which this author feels is relevant even today:

> As usual, when nobody knows anything concerning the problem, everybody forms his own theory and writes a book on this problem. And when the problem is understood, then it is no longer necessary to write books, but only to take two pages of the writing paper, sufficient to solve such a problem without difficulties.

Just such a reality exists in the field of the so-called diagonal tension theory, described by Jack McCormac in his book, *Design of Reinforced Concrete*: "Despite all this work and all the resulting theories, no one has been able to provide a clear explanation of the failure mechanism involved."[1]

Further, this author believes that the most competent forum in the world on diagonal tension is the Joint Committee of the American Society of Civil Engineers (ASCE) and the American Concrete Institute (ACI) 326. The members of Committee 326 are the most eminent experts in this domain and in their report, "Shear and Diagonal Tension", dealing with the mechanism of diagonal failure in concrete elements, they have recommended further investigations:

> Distributions of shear and flexural stress over a cross section of reinforced concrete are not known. . . . It has been pointed out in this report that the classical procedures are questionable in their development, as well as misleading and sometimes unsafe in the application. Yet, the goals of a complete understanding and of a fully rational solution to the problem of computing diagonal tension strength have not been attained.
>
> It is again emphasized that the design procedures proposed are empirical because the fundamental nature of shear and diagonal tension strength is not yet clearly understood. Further basic research should be encouraged to determine the mechanism which results in shear failures of reinforced concrete members. With this knowledge it may then become possible to develop fully rational design procedures.[2]

This is the motivation behind the writing of this book, which contains the results of 30 years of research by the author.

The synthesis of the author's work is his direct response to the recommendation of Committee 326. The result is the clear explanation of the mechanism leading to the diagonal failure of reinforced concrete beams, using the recommended practical method (based on the real stress condition of the bent member) for the calculation of reinforcement against failure. In other words, there is no longer a problem.

The work itself is divided into six chapters, each of which is an independent entity (with a seventh chapter dealing only with combined conclusions for the previous six chapters). Such a concept allows the reader to understand each detail in a given chapter without requiring preliminary information to be obtained from other chapters in the book. Thus, it is possible to consider each chapter as an independent study and an autonomous whole. Here, the author gives a general review for each chapter, enabling one to see what it is about, and what is new and original in this work.

Chapter 2, dealing with the mechanism of diagonal failure of a reinforced concrete beam, discusses experiments on wood laminated beams where the existence of internal active and internal resisting forces is proved in a physical way. The practical application of the new method is presented here by parallel calculations for diagonal failure using both the classical method and the new method. It is illustrated clearly how, with a complete understanding of the failure mechanism, the appearance of diagonal cracks can be prevented. The domain of the oscillations of the cracking angle at the diagonal failure is further elaborated in detail, pointing out, in particular, that when additional loads are being put on the beam, they must be transferred by special stirrups into the compression zone. At the conclusion of Chapter 2, the error of the classical theory on diagonal tension is clearly demonstrated. Namely, if such tension existed, it would actually be parallel to diagonal cracks and to the diagonal failure, itself.

The appendix deals with an analysis of internal active and internal resisting forces and is incorporated at the end of the book. Quotations are given from major university textbooks concerning the existence of internal active and internal resisting forces. Therefore, when the existence of these forces is accepted, the author's contribution will be accepted as well. This means that the author did not discover the existence of internal active and internal resisting forces – they were already known in technical literature. *The author's real contribution consists of the rational application of the existing forces and in the theoretical–practical demonstration that explains the following concepts and phenomena*:

1 The explanation, that the classical theory of diagonal tension and diagonal failure of beams of homogeneous or nonhomogeneous materials is only a hypothesis, unproven to this day.[3] This classical theory is disproved in this book.
2 The explanation of the material failure mechanism caused by fatigue in elements exposed to vibrations.

3 The error in the theory of the truss analogy, used as the basis for calculations for reinforced concrete structures.

4 The explanation that the diagram of shear forces of the beam over two supports is fundamentally different from the diagram of the fixed beam, for every kind of material (wood, steel, concrete, reinforced concrete, stone).

5 The application of the new theory concerning triangular reinforcing in a situation where seismic forces are involved, enabling a designer to obtain a much higher safety factor with the same quantity of reinforcement.

6 A new method of designing horizontal concrete membranes, for buildings and bridges, which resist seismic forces.

7 A new, very clear explanation of why cracks due to "diagonal tension", appear at a distance equal to the depth of the beam itself.

8 A fundamentally new explanation of diagonal failure of the beam, independent of the kind of material subjected to bending.

9 The equilibrium of a free body cut out from the bent beam can be proved algebraically by the application of the new method, based on the existence of internal active and internal resisting forces.

10 It follows, as the consequence of the existence of internal active and internal resisting forces, that there must exist, as well, two diagrams of shear forces: the diagram of active shear forces and the diagram of resisting shear forces.

11 Under the current state of knowledge and understanding of engineering mechanics, Newton's first law is for translatory equilibrium, but not for rotational equilibrium, and is therefore not applicable in the case of the equilibrium of a free body cut out of a bent beam. Furthermore, the existence of internal active and internal resisting forces in the bent beam is mentioned neither in the first nor in the third of Newton's laws. So, based on existing knowledge, this is quite a new concept for physics and for engineering mechanics as well.

Consequently, this new law of physics and its practical application in engineering science is explained for the first time! This law, as such, will revolutionize the thinking in the structural design of any flexural member; from any material, for any type of structure.

Chapter 3 deals with the error of the truss beam analogy for the reinforced concrete beam, and very clearly illustrates the faults in such a theory. Because this hypothesis is the basis for calculations involving reinforced beams, it is necessary to prove that such a hypothesis is fundamentally wrong (as well as the hypothesis on the pure shear existence in a bent beam). This is because a truss cannot be formed under any conditions in a cracked or uncracked beam loaded in bending. The principle of each truss is the formation of compression and tensile members. How can compression members be formed in the tensile zone? It is obvious that as long as the tensile zone exists in the bent beam, the compression member can never be formed. In addition, as the neutral axis moves up (not down) when it touches the top of the compression zone, an arch (or vault) can be formed but definitely not the truss. The truss, with compression members, can be formed only if the neutral plane moves downward and touches the bottom of the tensile zone. In this case, the formation of tensile members is not possible in the compression zone.

Chapter 4 deals with the failure of a beam due to material fatigue caused by vibrations and explains that the inflection points are the nucleus for crack formation and for failure due to material fatigue. This is a consequence of the law of physics that the inflection points are the static supports of the vibrating element while external forces, including physical supports, are factors conditioning the formation and location of the inflection points themselves. Besides cracks, conditioned by external forces by their physical actions in static supports, the inflection planes are the focus of additional stress accumulation, because of the fact that no building material, after unloading, returns to its original shape and position. When the compression zone becomes tensile, passing from a positive to a negative moment due to vibration, its extreme fibers are shorter and thus additional tensions are created in the previously shortened fibers, since they must have the same length as before the vibration. The connection with the other beams in the inflection planes does not allow this beam to remain shorter. As in the case of the inflection planes, where the connected fibers are shorter on one side of the inflection and longer on the other, it is normal in these inflection planes that the first additional tensile stresses appear only here.

In particular, it is explained that three different kinds of failure can occur at the inflection points. Each can cause the failure of the vibrating element, either separately, or in mutual combination:

1   Failure by pure shear through the inflection plane.
2   Failure by diagonal shear, where external supports, as forces, cause shear or separation of their parts of the beam at an angle of 45° in the direction of their action. Concurrently, the external loads, as forces, move their part of the beam to slide in the opposite direction. The cracks get unified with the cracks caused by the shear forces acting in the compression zone. This accelerates the failure because of the combination of compression forces and vertical shear forces whose resultants are acting parallel to the crack (which has occurred) but opposite to each other.
3   Failure near the inflection point (at a distance approximately equal to the depth of the beam, where the internal compression and tensile forces are equal to the vertical shear forces); the point at which, by combination of internal active compression and tensile forces with internal active vertical shear forces, punch shear is created leading to the failure.

Chapter 5, on triangular reinforcement of a shear wall (resisting much greater forces, than the wall reinforced according to the classical method), is a practical application of the new method of design for shear walls. Here, the difference between the beam on two supports and the shear wall is explained. It can be seen that these two elements function in different ways, so that statically they must be treated according to the function that they are performing in the given construction. Classically, the cantilever wall is treated as a part of the simple beam, the fixing of the cantilever being achieved where the vertical shear stresses change sign. The support of the beam is the reaction acting downward and the load acting upward, as shown in Figure 2.10. In other words, this is the part of the simple beam vertically rotated by 180 . The main difference here is that: for shear walls, one diagonal is exposed to elongation, and the other to shortening, while for a simple beam, such deformations of the diagonals do not exist. It is emphasized that the elongation of the diagonals is perfectly controllable by the diagonal reinforcement, where forces act along the reinforcement, so that each component of the shear force can be controlled by corresponding reinforcement. This was not possible with the classical method where, by adding horizontal reinforcement, resistance to diagonal failure cannot increase in low shear walls.

Chapter 6, dealing with the mechanism of horizontal membrane deformations, explains that if lateral forces are acting from two contiguous sides on a rectangular horizontal membrane, then, according to the classical theory on diagonal tension, the so-called diagonal tension is acting in two quadrants, while there is torsion in the other two quadrants. Applying our method, the membrane is simultaneously loaded to pure shear and to flexural bending under the action of lateral forces. By the application of triangular reinforcement, this problem is ideally solved, since external forces will act along the reinforcing bars, and the location and exploitation of the reinforcement is the most rational in comparison with stirrups (horizontal reinforcement) where they are idle before they are crossed by the cracking.

Chapter 7, a combined conclusion for all six chapters, explains that in any flexural bent member, two fundamentally different groups of internal forces are simultaneously created. The first group of internal forces is known as internal resisting forces (or reactions) to external active forces, also known as action and reaction. The second group of forces is the internal active and internal resistive forces of compression, tension, horizontal and vertical shear. The combination of these internal active forces causes diagonal cracking and diagonal failure of a loaded beam.

Also, this chapter summarizes that, in flexural bending, we distinguish two critical cross sections: one caused by critical moment and the second caused by a combination of internal active compressive, tensile, horizontal and vertical shear forces at the location where these forces become equal in magnitude in a loaded member. This cross section, where diagonal cracking and failure occurs, is located at a distance from the support which is approximately equal to the depth of the member.

Finally, this chapter explains how a safer design is possible for an inverted concrete T beam and for ductile steel framed structures.

References

1. McCormac, J. C. (1986) *Design of Reinforced Concrete*, 2nd edn, Harper and Row, New York, pp. 190, 191.
2. ACI-ASCE Committee 326, "Shear and Diagonal Tension", Proceedings, *ACI*, Vol. 59, January–February–March, 1962, pp. 3, 7, 18, 21.
3. Wang, C. and Salmon, C. (1965) *Reinforced Concrete Design*, International Textbook Company, Scranton, Pennsylvania, p. 63.

Chapter 2

Mechanism of diagonal failure
of a reinforced concrete beam

2.1 A brief overview of the problem

In this chapter, we show clearly that measurement of the strains of diagonal tension in a flexural member cannot be made directly. Rather, they have been determined indirectly from the principal strains by trigonometry. There appears to be no reliable physical evidence to support the existence of diagonal tension. This chapter also gives clear evidence of the fact that diagonal tension (as suggested by the pioneers) is only an assumption. This is illustrated in Figure 2.4.

Further, this chapter explains that, in a flexural member, two fundamentally different types of forces are developed simultaneously:

1   External bending forces and their corresponding internal resisting forces act as any push–pull force does, the external action causes an internal reaction (Figure 2.7(a)).
2   Internal active (compression, tensile, horizontal and vertical shear) forces and their corresponding internal resisting forces form a couple (active and resisting) located in the same plane and same line, equal in magnitude, but oppositely oriented.

Physical evidence for the existence of such forces has been shown by the splitting (separation through a neutral plane into two pieces) of a wooden beam exposed to flexural bending. Forces which cause such splitting are internal active forces, located at the surface of the sliding plane, oriented toward the sliding direction of such plane or portion of the beam. Yet, the forces which delay the appearance of splitting at a lower load, are internal resisting forces, located on the same surface of the sliding plane, oriented oppositely to the sliding direction of such plane or portion of the beam. They are cohesive as a material which resists the splitting of the beam into two pieces. Further, at any sliding area, two planes exist infinitely close to each other. In each plane, internal active and internal resisting sliding shear stresses exist simultaneously, as illustrated in Figure 2.7(a).

This chapter also explains that, as a result of a combination of internal active compression and tensile forces with vertical internal active shear forces, diagonal cracking and diagonal failure occur. As a result of the existence of internal active and resisting forces, the equilibrium of a free body becomes algebraically provable.

2.2 Technical analysis

The existing misunderstanding of the previously unexplained diagonal cracking phenomenon of concrete members can be very well illustrated with the following quote from the Joint Committee (ACI–ASCE) 326:

> It has been pointed out in this report that the classical procedures are questionable in their development, as well as misleading and sometimes unsafe in their application. Yet, the goals of a complete understanding and of a fully rational solution to the problem of computing diagonal tension strength have not been attained.[1]

The fact that vertical stirrups remain idle (free of any tension) during bending of a flexural member before they are crossed by the crack itself is supported by five different references.[2] This author knows of no reasonable explanation of why and how this can be so! The statement of Joint Committee 426: "Prior to flexural cracking all the shear is carried by the uncracked concrete,"[3] cannot be supported by common sense. It is impossible that, at a given cross section through the vertical stirrup, only the particles of concrete would be loaded by "shear," while the same stirrup, glued by cement gel to the same particles, would be free of any load before it is struck by diagonal cracking. Some small stresses would be introduced in the stirrups as a result of Poisson's ratio, but these stresses are much lower than the stresses capable of causing diagonal cracking in a concrete mass.

There is no reasonable explanation why "for the specimens with height-to-horizontal length ratio of half and less, it was found that horizontal wall reinforcement did not contribute to shear strength."[4] Our theory provides the explanation for why this is so.[5] There is no explanation as to why in the flexural members:

a   "In both cases, failure occurred with a sudden extension of the crack toward the point of loading;"[6] ... "Cracks in the pure moment section were vertical, while cracks in the shear span curved toward the point of the applied load as soon as they entered the area between level of the tension reinforcement and the neutral axis" (Figure 3 in reference 6).

b   In contrast with a concentrated load, a member exposed to overloading by a uniform load will show punch shear cracks which lead to a fracture at the support: "Failures which have occurred in American large panel structures indicate that the weakest point of the section under debris loading is the shear strength of the floor at the support."[7,8] This statement cannot be explained by the classical theory of diagonal tension.

There is no reasonable explanation to clarify why "Many equations currently in use have little relevance to the actual behavior of the member at the stage of the crack development;"[9] ... "In other words, the classical design procedure does not correlate well with the test results."[1] Also, "The test results obtained from several hundred beams have shown that no direct relationship exists between "shear strength" and shear force V."[10]

It has never been understood why a T (tee) beam has about 20 percent greater resistance against diagonal "tension" cracking than a beam with a rectangular cross section.[3] Nor does an explanation exist as to why "at the interior support of a beam the diagonal cracks, instead of being parallel, tend to radiate from the compressed zone at the load point."[9] The punch shear theory put forward by us explains these phenomena very well.

The mechanism of diagonal cracking has not been understood: "The actual mechanics of the formation of diagonal-tension cracks and the manner in which the steel assumes the load is still not completely understood,"[11] Professor Priestley, in his article, "Myths and Fallacies in Earthquake Engineering – Conflicts between Design and Reality" (ACI SP 157, published 1993) declared: "Shear design of reinforced concrete is so full of myths, fallacies and contradictions that it is difficult to know where to begin in an examination of current design. Perhaps the basic myth, and central to our inconsistencies, in shear design is the issue of shear itself. It has been argued that we tie ourselves into intellectual knots by separating flexure and shear, and considering them essentially independent entities." ... "Despite all this work and all the resulting theories, no one has been able to provide a clear explanation of the failure mechanism involved."[12] Yet, this book (and previous publications)[8,13,14] offers just that; a clear explanation. Unbridgeable differences between diagonal tension failure caused by torsion or pure shear and punch shear failure are illustrated in references 2, 15–17. The facts quoted above, about the misunderstanding of diagonal tension have been the main motivation for preparing the following discussion, with a view to resolving the mystery. This book will investigate diagonal tension due to stresses caused by bending and will only, from time to time, compare them with stresses, as the existing theory suggested, caused by "pure shear."

2.3   Comment on the existing theory of shear and diagonal tension

There are a minimum of seven obstacles in existing structural theory which are primarily responsible for the misunderstanding of the mechanism of diagonal cracking in a flexural concrete member. These are listed in the following sections.

2.3.1   Direction of shear forces are applied as we need them and not as they exist

(a)   Concerning horizontal shear stresses (forces)
In the book *Advanced Mechanics of Material*, 1961 edition (Wiley), Seely and Smith showed (for a flexurally bent beam on two supports) horizontal shear stresses

above the neutral plane oriented toward the supports and below the neutral plane oriented toward the critical cross section as illustrated in Figure 2.1. Fibers of a wooden beam on two supports will separate from each other by sliding against each other precisely as illustrated in Figure 2.1. But when sliding already occurs in a simply supported wooden beam, the same authors show stresses on a unit element (located in a neutral plane), oriented oppositely, only to prove that such separation (sliding) is a result of diagonal tension as illustrated in Figure 2.2. To quote, "In a beam made of timber, however, the stresses σ and τ at the section of maximum bending moment may not be significant because the grain of the wood may be such as to make wood very weak in shear on a longitudinal plane, and hence the horizontal or longitudinal shearing stress τ_H (shown in Figure 3b) may be the significant stress in the failure; a failure of a wooden beam by longitudinal shear is shown in Figure 5" (pages 26 and 27 in their book). What irony, that to

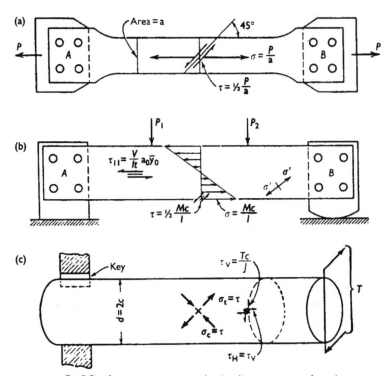

FIG. 3 Significant stresses as given by the elementary stress formulas.

Figure 2.1   Seely and Smith show, in their Figure 3b, the direction of horizontal shear stresses, as they exist and act in any flexural bent member. This is in full agreement with Saliger's presentation of shear stresses, as shown in Figure 2.5(b).

Source: Seely and Smith, Advanced Mechanics of Material, 1961 edition (Wiley).

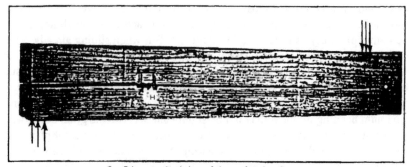

FIG. 5 Longitudinal shear failure of timber beam.

Figure 2.2 In order to explain that the longitudinal failure of the timber beam above is caused by the so-called diagonal tension, Seely and Smith were forced to apply the direction of shear stresses opposite to the stresses shown in Figure 2.1. By so doing, they contradicted themselves and, at the same time, violated a law of physics.

Source: Seely and Smith, *Advanced Mechanics of Material,* 1961 edition (Wiley).

prove what never existed, we are closing our eyes to the truth of the fallacy of the diagonal tension theory!

These two figures (Figures 2.1 and 2.2) are indeed very fine examples of the fact that we are applying orientations (directions) of sliding shear as we need them and definitely not as they really are in the stressed body of a flexurally bent member.

(b)   Concerning vertical shear stresses (forces)

For a portion of a given beam, at a given cross section, the same vertical shear force V has been said to act *upwardly* in order to prove the accuracy of the truss analogy theory, and to act *downwardly* in order to prove the existence of diagonal tension in a flexural member. In other words, the same active force V can act (in the same flexurally bent member and the same cross section) in opposite directions as we need it, depending on which theory we want to prove. (See Figure 1b of reference 18, or compare vertical shear force V for the truss analogy theory with the corresponding shear force V used to prove the diagonal tension theory, illustrated in Figures 2.3 and 2.4.)

2.3.2 Applied directions of vertical shear stresses contradictory to their natural directions

Horizontal and vertical active shear stresses have been shown by Professor Saliger (1927)[19] as *acting in one direction* (as they should naturally act). But, in order to prove the existence of diagonal tension, the same active shear stresses at the same location in a beam are shown in standard engineering textbooks as acting *in the*

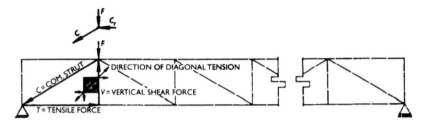

Figure 2.3 This figure illustrates the fact that if we attach a unit element (block) to the vertical shear force V, developed as a part of truss analogy theory, it appears that diagonal tension is parallel to diagonal cracking.

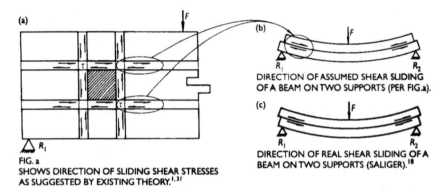

FIG. a
SHOWS DIRECTION OF SLIDING SHEAR STRESSES
AS SUGGESTED BY EXISTING THEORY.[1,31]

Figure 2.4 As illustrated in (a), the direction of shear stresses (forces), based on existing theory clearly suggests horizontal sliding of a flexural member to be as shown in (b) (which is opposite to the direction of real sliding shear forces as shown in (c)). If such sliding shear stresses (forces) somehow could exist as shown in (a) and (b), then they will be neutralized (canceled) by oppositely oriented real sliding shear stresses (forces) as shown in (c); therefore the only explanation of diagonal cracking could be as suggested in Figure 2.8(c).

opposite direction (see Figures 2.5 and 2.7). Vertical shear stresses τ_v (as shown by Saliger in Figure 2.5(a)) act upwardly at the left-hand side. Yet, they will act upwardly on the surface of any unit element (block) as long as the right surface (plane) of a block is a part of the plane of the left cross section shown. Separation of a block from the given cross section cannot change the direction of shear stresses τ_v (see block shown between Figures 2.5(a) and (b)).

2.3.3 Pure bending proves the existence of internal resisting forces which are used to determine active shear stresses

For equilibrium of a portion of the beam which is exposed to pure bending (where bending forces are parallel to the longitudinal axis), resisting compression and resisting tensile forces C_r and T_r (as the only available forces to maintain balance) have been applied (Figure 2.8(a)). Yet, in order to prove the truss analogy theory in an ordinary bent member (where bending forces are perpendicular to the longitudinal axis), the resisting forces (C_r and T_r) are converted in textbooks to be active compression and tensile forces (C and T) and these forces are applied for equilibrium of a given portion of the beam (see Figures 2.7(b) and 2.8(b)). Without such conversion, the existence of the compression strut of the alleged truss cannot be proved. How can this conversion be possible? How is it that for pure bending, where vertical shear is lacking, resisting forces (C_r and T_r) must be applied; while for the same portion exposed to ordinary bending, active compression and tensile forces (C and T) must be applied as a result of the presence of vertical shear forces V? Can the presence of vertical shear forces convert resisting forces (C_r and T_r) to active forces (C and T)? If active forces (C and T) could create a resisting moment M_r, what should resisting forces (C_r and T_r) do? Yet, these resisting compressive stresses (forces) are applied in Mohr's circle, in order to determine active shear stresses in a bent member.

2.3.4   The + or − sign does not have any relation with the origin or direction of shear forces

Usually in technical literature, a sign is used to show the direction of movement of some element (electricity, temperature, moment, etc.). This should also be the case with signs denoting shear: the direction of shear stresses should go up or down in correlation with the plus or minus sign. Unfortunately, the sign convention means only that a given *vertical shear stress* (which could be real or false) is located at one side or the other of the critical cross section of the given beam. The sign does not show or define direction of action of vertical shear stresses (forces). So the same shear forces for the same moment diagram (positive or negative) with the same direction of action would be called "positive" if they are located on one side, and "negative" if on the other side of the critical cross section of the given simply supported beam. Even in a shear diagram for a fixed end beam, the same sign (plus or minus) is used for vertical shear forces acting in opposite directions: the bending forces for the cantilever portion act upward while those for the simple beam act downward.

Furthermore, if vertical shear stresses at the left side of a simply supported beam are oriented in a counter-clockwise direction then, in this area, they would be negative instead of positive, as is the case with shear stresses caused by bending and shown to be so by Saliger[19] (Figure 2.5(a)).

Because positive or negative shears, in the nature of a flexural member, do not exist nor do any changes of direction of vertical shear stress action in a simple beam, the sign by itself proves or disproves nothing. If such direction of shear stresses does exist in a flexural member, then "the vertical shearing force acting at a cross section of a beam will be considered to be positive while it points in a direction as to produce a clockwise moment,"[20] ... "the shear stress τ on any face of the element will be considered positive, when it has a clockwise moment with respect to a center inside the element."[21] Evidently, as far as the direction of shear stresses is concerned, the sign is a source of confusion leading to misunderstanding of the concept of shear and diagonal tension.

There is no math or authority that can negate the natural direction of *sliding* shear forces (stresses). The portion above the neutral plane of the beam wants to slide to the supports, while the corresponding upper surface of a lower portion tends to slide towards the center of the beam. So if we place a unit element directly at the left or right side of Saliger's vertical shear stresses in Figure 2.5(a) (directly above or below the horizontal shear stresses), diagonal tension should occur parallel to the diagonal crack and yet τ_v would be negative shear and τ_h, positive shear. This is probably the best illustration that something is fundamentally wrong with the classical concept of diagonal tension; because the information gathered through seeing the direction of sliding is much more accurate than any information got by the assumption that, from the principal strains, shear stresses are determined via trigonometry because, "accurate measurement of the shearing strain is found to be very difficult. It is easier and more accurate to measure ... the principal strains ... and convert them in shearing strain."[21]

2.3.5 A shear diagram is probably the biggest obstacle to understanding the shear phenomenon

From a scientific point of view, it is possible that the classical shear diagram has become one of the worst culprits in preventing our understanding of the direction of shear stresses (forces) in a flexural member. Even though the direction of bending shear forces in a fixed-end beam is fundamentally different from the direction of the same forces in a simply supported beam, both are represented by the same shear diagram. How can a shear diagram representing two negative moments and one positive moment be identical to the shear diagram for a simply supported beam? How can a shear diagram be useful without showing the direction of action of shear forces? How can we know the direction of shear stresses if we do not have the slightest idea of the direction of shear forces? In fact, we developed Figures 2.10 and 2.11 primarily for the structural engineering profession to see that the direction of vertical shear forces must follow the bending line of the beam. With such knowledge, it is easy to see the fallacy of diagonal tension by applying the corresponding vertical shear forces at the cross section located in the plane of one surface of a unit element (block), as has been shown in Figures 2.5(a)–(c).

Stated another way, a fixed-end beam and a simple beam are each represented by the same shear diagram in classical theory even though the fixed-end beam is composed of two cantilevers, and a simple beam which is located between two inflection points (PI). The simple beam is composed of only one beam. In other words, a fixed-end beam has one concave and two convex bending lines while a simple beam has only a concave bending line. Furthermore, the bending shear forces for a simple beam are oriented only in one direction (say, downward) while the bending shear forces in a fixed-end beam are oriented in opposite directions: the bending forces (supports) of the cantilevers are oriented upward while the bending forces of the simple beam (between the two PIs) are oriented downward. Finally, a simple beam does not have any PI, while a fixed-end beam has two PIs with shear forces oppositely oriented at the PIs.

One cannot understand the phenomenon of shear if one uses the same shear diagram to represent the bending shear forces for a fixed-end beam and for a simple beam and one does not have the slightest idea of the direction of shear forces in the shear diagram for any type of bending.

2.3.6 The assumption that pure shear exists in flexural bending makes the understanding of the shear phenomenon impossible

Undoubtedly, the assumption of the existence of pure shear (which eventually should cause diagonal tension in a flexural member) made a great contribution to the misunderstanding of diagonal failure in a concrete member. There is no logic behind this assumption nor has it been made with a reasonable understanding of the mechanism of diagonal cracking. Rather, the concept that, for diagonal cracking to occur, there must be forces acting perpendicularly to the crack has given rise to this assumption. While unable to visualize the existence of any other forces capable of causing such cracking, the pioneers in structural theory assumed the possible existence of pure shear which could generate diagonal tension. The concept that a bent member, by itself, could originate from punch shear forces (caused by a combination of a vertical shear force with a tensile force) was totally strange to them, so the only possible explanation could be the existence of pure shear.

This "pure shear" theory, in its simplest form, can be illustrated by this quote from a textbook: "During the years since the early 1900s until the 1963 code was issued, the rational philosophy was to reason that in regions where normal stress was low or could not be counted on, a case of pure shear was assumed to exist."[22]

Unfortunately, up to the present, nobody ever dared to challenge such an assumption to see whether the assumed active shear stresses could coexist with active shear stresses caused by bending. Yet, sliding shear stresses caused by bending could not be ignored because their existence and their cause is a real and natural fact. Anyone could understand such sliding stresses if it was understood that the sliding phenomenon of a beam on two supports was as illustrated in Figure 2.5. Such stresses have nothing to do with being wrong or right because they are part of the bending phenomenon itself.

It appears that the assumed direction of active shear stresses (according to the classical theory) is oppositely oriented to the direction of real sliding shear stresses (caused by the bending phenomenon). This has been illustrated by Professor Saliger (1927) and shown in Figure 2.5. So, in considering active shear stresses caused by the bending phenomenon and assumed active shear stresses shown on a unit element in any textbook, there appears to be no correlation. For that very simple reason, the classical theory of diagonal tension was not understood because real sliding shear stresses in a bent member are oriented in one direction while the so-called sliding shear stresses in the classical theory are oriented in the opposite direction (see Figure 2.4). For that same reason, "many equations currently in use have little relevance to the actual behavior of the member at the stage of crack development,"[9] that is, there is no correlation between the equations and the laws of nature.

Finally, if we assume that the classical theory is correct, then in any bent member, three types of diagonal tension should exist simultaneously:

1   First, perpendicular to diagonal cracking (due to the existence of pure shear) as the classical theory assumed.
2   Second, parallel to diagonal cracking, caused by sliding shear stresses due to the bending phenomenon (see Figure 2.5).
3   Third, caused by punch shear forces created by a combination of vertical shear forces, with tensile forces as shown in Figure 2.8(c).

But, in such cases, diagonal tension perpendicular to diagonal cracking will be neutralized (canceled) by diagonal tension parallel to diagonal cracking as a result of the opposite direction of action of active shear stresses. At the end of such cancellation there will remain active and real "diagonal tension" caused by punch shear forces V_n as shown in Figure 2.8(c).

2.3.7 Principal stress trajectories do not have any relation with the origin of shear stresses

Principal stress trajectories represent algebraic functions expressed geometrically as lines. Therefore, they can be useful only if the given equation represents a given phenomenon. But, because existing theory is based on the premise that "a case of pure shear was assumed to exist,"[22,1] (from which equations for principal stress trajectories are developed), the question arises: what does this mean to us? The answer is: the direction of active shear stresses is assumed to be such as they are shown in any textbook. Now a decisive question arises: Does the direction of these stresses coincide with the direction of shear stresses caused by the bending phenomenon itself? Tragically, the answer is no. Shear stresses caused by the bending phenomenon are oriented oppositely to those supposed by our pioneers. This fact could be visualized by observing the sliding of a beam on two supports, as illustrated in Figure 2.5, or by comparing the assumed direction of shear stresses with the direction of the shear stresses presented by Professor Saliger[19] (as shown in Figure 2.4). As a result of the opposite directions of the assumed and real shear stresses (caused by bending), the trajectories cannot prove or disprove anything.

Yet, given that equations (as a mathematical concept) are very correct (even though they do not represent the real stress condition of a flexural member), and

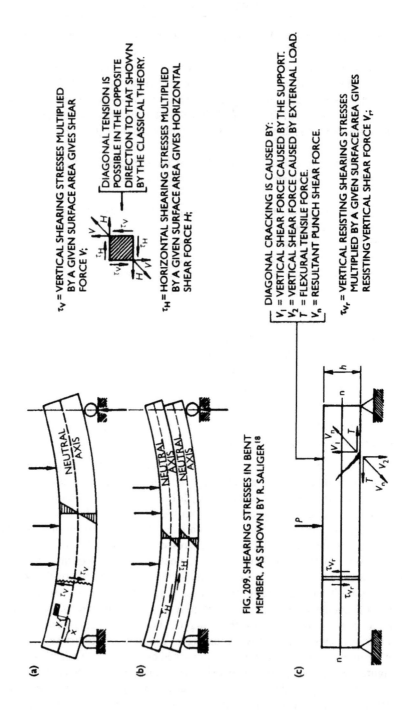

τ_V = VERTICAL SHEARING STRESSES MULTIPLIED BY A GIVEN SURFACE AREA GIVES SHEAR FORCE V:

DIAGONAL TENSION IS POSSIBLE IN THE OPPOSITE DIRECTION TO THAT SHOWN BY THE CLASSICAL THEORY.

τ_H = HORIZONTAL SHEARING STRESSES MULTIPLIED BY A GIVEN SURFACE AREA GIVES HORIZONTAL SHEAR FORCE H:

DIAGONAL CRACKING IS CAUSED BY:
V_1 = VERTICAL SHEAR FORCE CAUSED BY THE SUPPORT.
V_2 = VERTICAL SHEAR FORCE CAUSED BY EXTERNAL LOAD.
T = FLEXURAL TENSILE FORCE.
V_n = RESULTANT PUNCH SHEAR FORCE.

τ_{V_r} = VERTICAL RESISTING SHEARING STRESSES MULTIPLIED BY A GIVEN SURFACE AREA GIVES RESISTING VERTICAL SHEAR FORCE V_r:

FIG. 209. SHEARING STRESSES IN BENT MEMBER. AS SHOWN BY R. SALIGER[18]

τ_{Hr} = HORIZONTAL RESISTING SHEARING STRESSES MULTIPLIED BY A GIVEN SURFACE AREA GIVES RESISTING HORIZONTAL SHEAR FORCE H_r;

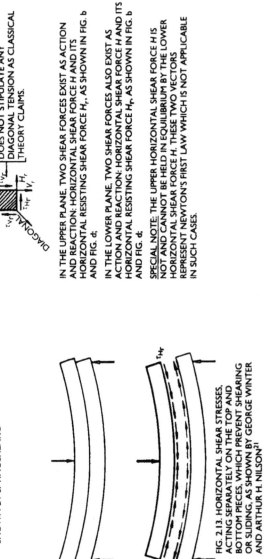

EQUILIBRIUM OF A UNIT ELEMENT DOES NOT STIPULATE ANY DIAGONAL TENSION AS CLASSICAL THEORY CLAIMS.

IN THE UPPER PLANE, TWO SHEAR FORCES EXIST AS ACTION AND REACTION: HORIZONTAL SHEAR FORCE H AND ITS HORIZONTAL RESISTING SHEAR FORCE H_r, AS SHOWN IN FIG. b AND FIG. d;

IN THE LOWER PLANE, TWO SHEAR FORCES ALSO EXIST AS ACTION AND REACTION: HORIZONTAL SHEAR FORCE H AND ITS HORIZONTAL RESISTING SHEAR FORCE H_r, AS SHOWN IN FIG. b AND FIG. d;

SPECIAL NOTE: THE UPPER HORIZONTAL SHEAR FORCE H IS NOT AND CANNOT BE HELD IN EQUILIBRIUM BY THE LOWER HORIZONTAL SHEAR FORCE H. THESE TWO VECTORS REPRESENT NEWTON'S FIRST LAW WHICH IS NOT APPLICABLE IN SUCH CASES.

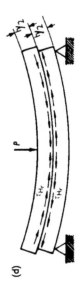

(d)

FIG. 159. HORIZONTAL RESISTING SHEARING STRESSES WHICH PREVENT SLIDING, AS SHOWN BY S. TIMOSHENKO[14]

(e)

(f)

FIG. 2.13. HORIZONTAL SHEAR STRESSES, ACTING SEPARATELY ON THE TOP AND BOTTOM PIECES, WHICH PREVENT SHEARING OR SLIDING, AS SHOWN BY GEORGE WINTER AND ARTHUR H. NILSON[21]

Figure 2.5 This figure illustrates four different types of shear forces: (1) Vertical shear force F; (2) vertical resisting shear force V_r;[25,37] (3) horizontal shear force H; and (4) horizontal resisting shear force H_r.

because we believe that everything which can be expressed mathematically should be correct, the concept of trajectories probably made an enormous impact on the never-to-be-understood phenomenon of diagonal cracking in a concrete member. Even in schools, trajectories are used to prove the "accuracy and precision" of the diagonal tension theory.

Additionally, in schools we have been taught that statics is a part of mathematics; so, as such, it is an exact science and nothing could be wrong with it. Evidently, the concept of the assumption of the possible existence of pure shear in flexural bending was not well known to our teachers. As a result of such philosophy, no one dared to challenge the accuracy of the diagonal tension theory. Therefore, it survives to this day.

2.4 Discussion

Until now there has been much confusion and misunderstanding concerning internal and external forces. In technical literature, the term "internal forces" has been understood exclusively as reactions to active forces. This is correct only for some cases of pure "push or pull" forces. For example, our bodies cause reaction (internal force) on the floor. But external bending forces in flexural bending phenomena cause internal active and internal resisting forces, in addition to causing external action and internal reaction. Horizontal sliding shear forces (stresses τ_H), as shown by Saliger,[19] and horizontal resisting sliding shear forces (stresses τ_{H_r}) as shown by Timoshenko[20] and Winter-Nilson,[23] clearly illustrate that in flexural bending, internal active and internal resisting forces do indeed exist in the same way as they exist in a twisting shaft or in torsion.

In order to simplify and to make the stress condition of a unit element (block) cut out from the flexurally bent member easily understandable, consider the following analogy. What was said of an atom eighty years ago in physics (smallest possible particle that could be obtained from any matter by physical or chemical analysis) could now be said for diagonal tension. The existing theory of diagonal tension must be "correct" because no one has proved it to be otherwise. This is just the way it was with the atom until somebody proved it to be otherwise. What an electron and proton were for an atom is now applicable to active and resisting forces (stresses) as they act on a free body or on a unit element. The law of physics is not a miracle, but simple common sense. If there exists an action, inevitably there must exist a reaction, or the free body will continue its movement in space forever. If there is no reaction to the force our body exerts on the floor, our body would continue to move towards the center of earth's gravity. Our weight causes a reaction in the floor of equal magnitude but opposite direction (here, upward), touching each other at the floor's surface. That is, action and reaction must touch each other and, by counteracting, make equilibrium.

Our weight is an excellent example of the case where the internal forces which arise are indeed a reaction to active external forces. That is a reason for the lack of understanding about the flexural bending phenomenon and consequently, diagonal tension. Therefore, a picture has already been formed in our mind, that internal forces must exclusively be a reaction to external forces. So any internal forces (including internal forces in a flexural bending) must only be the resisting forces – what else can they be? Also, this example of pure push–pull forces is fundamentally different from the phenomenon of flexural bending where bending, by its very nature, creates internal active forces and their corresponding internal resisting forces. Such internal active and resisting forces could be recognized by our senses simply by observing such sliding in a neutral plane of a bent member.

There is also a clear analogy here between the internal active and resisting forces of electrons and protons, or internal gravitational forces of any planet and our sun, making equilibrium-balance between each other. In both cases (electromagnetic or gravitational forces) equilibrium is based on internal active and their corresponding internal resisting forces, so the pure push–pull concept is a rather exceptional (specific) phenomenon in our universe. Consequently, internal active and their corresponding internal resisting forces are nothing new, as a concept, because they are responsible for the equilibrium of the universe itself. Also, the internal active and internal resisting forces in flexural bending are formed by the bending phenomenon itself.

The same must be true for any free body: if there exists any action on the body (or in the body), there must also exist its reaction, and vice versa: if there exists any reaction, it is evident that there must also exist its own action, or equilibrium (balance) is impossible. If a proton (as an internal force) were removed from the nucleus of the hydrogen atom, the electron would continue its movement in space forever. This could also be said for flexural bending of internal forces. So, whatever can be said for action and reaction for a positive proton and a negative electron (mainly caused by electromagnetism or opposite electrical charges), can be said for action and reaction for internal flexural bending forces. Any active force (stress) must have its own reaction or its movement would continue in space forever; it is irrelevant if such force is internal or external (force is force). This means that no one could speak of the equilibrium of any free body or unit element without very clearly seeing action and reaction. Yet, a false hypothesis never could lead to a real solution by showing active forces on the free body (C, T, V and H) without showing corresponding resisting forces (C_r, T_r, V_r and H_r) in order to establish real equilibrium (balance).

If there exists an active vertical shear force V on or in the body of any portion of a free body (unit element), there must also exist its own reaction V_r in or on the same body. Also, if any shear force V at any cross section of any free body exists, its own reaction must also exist at the same cross section or the shear force V would continue its movement in space or toward the center of gravity of Earth. Equilibrium is possible only after full recognition of active and corresponding resisting forces (stresses) acting in opposite directions to each other (at the same line or plane) in the same way that forces between an electron and proton act against each other making equilibrium of a hydrogen atom.

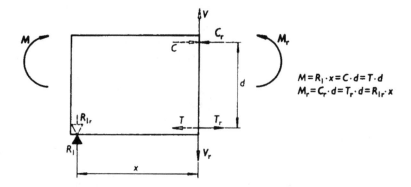

$$M = R_1 \cdot x = C \cdot d = T \cdot d$$
$$M_r = C_r \cdot d = T_r \cdot d = R_{1r} \cdot x$$

Figure 2.6 Equilibrium of a free body becomes algebraically provable by application of a new law: $\sum X = 0, \sum Y = 0$ because $C = C_r, V = V_r, R = R_{1r}$ and $M = M_r$.

This concept that the algebraic sum of vertical forces is to be equal to zero does not and cannot eliminate resisting forces. If the external active vertical forces are in equilibrium with their corresponding resisting vertical shear forces $V_r(\sum Y = 0)$, then automatically there is equilibrium of all vertical resisting forces (R_{1r} and all other vertical resisting forces) with their corresponding active vertical shear forces V, because the algebraic sum is zero (see Figure 2.6). Yet, the active vertical force R_1 has its own reaction (resisting force) R_{1r}, acting in an opposite direction (here, downward) as shown in Figure 2.6. At the same time, external concentrated force F developed its own resisting force (reaction) F_r, and R_2 developed its own resisting force (reaction) R_{2r} as shown in Figure 3.2. The vertical shear force V, caused by the support R_1, has the same direction as support (here, upward), while its reaction (resisting force) V_r has opposite direction (here downward). Since $R_1 = R_{1r}$ and $V = V_r$, the sum of vertical forces at the left portion is zero. Consequently, the left portion of the free body (in Figure 2.7(b)) must be in equilibrium where vertical forces are concerned (active force is in balance by its resisting force). If one shows, at a given cross section, vertical shear force V, then it must also show all resisting forces acting in the same body (at the right portion resisting forces F_r and R_{2r}) and the portion of the beam, as such, must be in equilibrium because $\sum Y = 0$.

Identical logic can be used for horizontal shear forces: horizontal shear forces H are in equilibrium by their resisting forces H_r. Because $H = H_r$, the sum of horizontal shear forces at the left portion of the free body is zero. So the free body (as far as horizontal shear forces are concerned) must be in equilibrium, as shown in Figures 5.1, 2.7(a) and (c). Also, the unit element (block) in Figure 2.7(c), must be in equilibrium by corresponding active, V, and resisting, V_r, shear forces.

Indeed, equilibrium of a block in Figure 2.7(c) is achieved by applying the concept of actions and corresponding reactions. By applying such logic, diagonal tension in flexural bending appears to be parallel to diagonal cracking as shown in

Figure 2.5. Such facts can only be interpreted to mean that diagonal tension cracking in a flexural member cannot be caused exclusively by vertical and horizontal shear forces, but rather by some completely different phenomenon, as illustrated in Figure 3.2 in Section 3.4.2.1 of Chapter 3.

Because the active forces and their corresponding resisting forces are of utmost importance to our theory, it should be reiterated that it is a decisive rule for any push or pull force, that no one force could be in equilibrium without some counter-active force ("a single, isolated force is therefore an impossibility").[24] Consequently, as shown by the existing structural theory, forces C and T, acting perpendicularly at a given cross section of any free body, cannot be in equilibrium even though compression force C is equal to tensile force T and that the sum of these two forces is zero. To repeat: they are *not in equilibrium* even though $\sum X = 0$ and $\sum F_X = 0$. They are in translational, but not in rotational equilibrium. Their internal moment is not their reaction but these forces could be replaced by their moment. At the same time, resisting moment M_r is in equilibrium with its own external moment M at a given cross section. Rotational equilibrium could be achieved by the moment of external forces for the given cross section.

Unfortunately, the horizontal shear forces H cannot be equilibrated by the corresponding moment, since there exist only three moments in the vertical plane of the bent beam. Thus, these forces are only in translational equilibrium as a couple, but not in rotational equilibrium. This means that the horizontal resisting shear forces $H_r(\tau_{hr})$, as the resistance of the material against shear, is the only possibility for equilibrating the horizontal shear forces $H(\tau_h)$.

Compression and tensile active and resisting forces in flexural bending can be found in almost every structural civil engineering textbook, where tensile forces are acting from the crack to the left and to the right, since the crack was created by tensile forces. This irrefutable law of nature is clearly illustrated in Figure 3.2, Section 3.4.2.1 of Chapter 3. It is obvious in this figure that tensile forces on separated parts of the free body are acting normally to this section, causing the crack. Since only active tensile stresses can cause cracks, it follows that tensile stresses, as shown in any textbook (acting out of the section), can only be resisting tensile stresses (forces). We have indicated these forces by T_r, where the subscript "r" denotes "reaction". It should be emphasized that the clear and distinct differences between active and resisting internal forces in a bent beam are extremely important for the comprehension and understanding of the diagonal failure of a bent concrete member.

In summary, it should be repeated that we have not invented or discovered the existence of internal active and internal resisting forces in a bent member. The author has drawn the scientists' attention to their existence, and explained why the diagonal failure occurs. The existence of a unique physical law concerning the bent elements has been indicated. *This unique law establishes that, only in flexurally bent members, internal active and internal resisting forces of compression, tension and shear occur, and that the combined action of internal active forces causes the diagonal cracks and the diagonal failure of concrete*

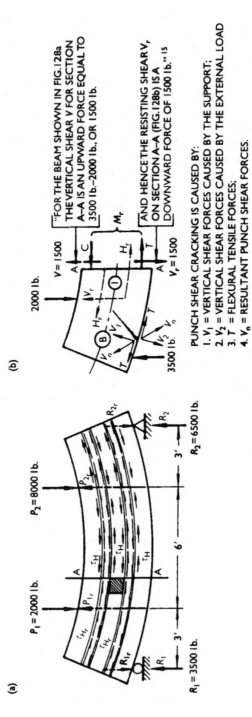

(a)

$P_1 = 2000$ lb. $P_2 = 8000$ lb.

$R_1 = 3500$ lb. $R_2 = 6500$ lb.

(b)

2000 lb.

"FOR THE BEAM SHOWN IN FIG.128a THE VERTICAL SHEAR V FOR SECTION A–A IS AN UPWARD FORCE EQUAL TO 3500 lb.–2000 lb., OR 1500 lb.

AND HENCE THE RESISTING SHEAR V_r ON SECTION A–A (FIG.128b) IS A DOWNWARD FORCE OF 1500 lb." [15]

$V = 1500$

$V_r = 1500$

M_r

3500 lb.

PUNCH SHEAR CRACKING IS CAUSED BY:
1. V_1 = VERTICAL SHEAR FORCES CAUSED BY THE SUPPORT;
2. V_2 = VERTICAL SHEAR FORCES CAUSED BY THE EXTERNAL LOAD
3. T = FLEXURAL TENSILE FORCES;
4. V_n = RESULTANT PUNCH SHEAR FORCES.

(FIG.128 INTERNAL FORCES AT ANY SECTION OF BEAM[15])

(c)

(d)

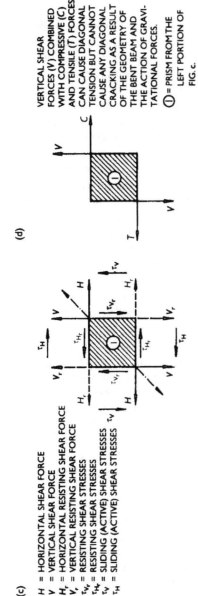

H = HORIZONTAL SHEAR FORCE
V = VERTICAL SHEAR FORCE
H_r = HORIZONTAL RESISTING SHEAR FORCE
V_r = VERTICAL RESISTING SHEAR FORCE
τ_{V_r} = RESISTING SHEAR STRESSES
τ_{H_r} = RESISTING SHEAR STRESSES
τ_V = SLIDING (ACTIVE) SHEAR STRESSES
τ_H = SLIDING (ACTIVE) SHEAR STRESSES

VERTICAL SHEAR FORCES (V) COMBINED WITH COMPRESSIVE (C) AND TENSILE (T) FORCES CAN CAUSE DIAGONAL TENSION BUT CANNOT CAUSE ANY DIAGONAL CRACKING AS A RESULT OF THE GEOMETRY OF THE BENT BEAM AND THE ACTION OF GRAVITATIONAL FORCES.

① = PRISM FROM THE LEFT PORTION OF FIG. c.

NOTES:

1. EQUILIBRIUM OF ANY FREE BODY, INCLUDING AN INFINITESIMALLY SMALL PRISM, IS BASED ON NEWTON'S THIRD LAW OR THE LAW OF ACTION AND REACTION: TO STOP THE MOVEMENT OF ANY FREE BODY ITS REACTION AS RESISTANCE OF MATERIAL TO SUCH MOVEMENT MUST BE USED, (SEE FIG. b AND c).

2. TO EXPLAIN THE ORIGIN OF DIAGONAL TENSION, HORIZONTAL SHEAR FORCES H AND VERTICAL SHEAR FORCES V MUST BE USED. BUT IN SUCH A CASE, THE DIAGONAL TENSION SHOULD OCCUR IN THE DIAGONAL *OPPOSITE* TO THE CLASSICAL CONCEPT AS SHOWN IN FIG. c.

3. IN FIG.a "HORIZONTAL SHEAR STRESSES TO PREVENT SLIDING" τ_{H_r} ARE SHOWN AS THEY ARE EXPLAINED BY TIMOSHENKO (16–FIG.159) AND AS "HORIZONTAL SHEAR STRESSES WHICH PREVENT SHEARING OR SLIDING" BY G. WINTER AND A.H. NILSON (23–FIG.2.13).

4. IN FIG.c ARE SHOWN ACTIVE HORIZONTAL (τ_H) AND VERTICAL (τ_V) SHEAR STRESSES WHICH ARE CAUSING SLIDING-SHEARING OF THE FLEXURALLY BENT MEMBER AS THEY ARE PRESENTED BY SALIGER (18, FIG. 209). THESE AND SUCH STRESSES (τ_V AND τ_H) CAN ONLY CAUSE EVENTUAL DIAGONAL TENSION AS SHOWN IN FIG. c.

5. V: VERTICAL SHEAR FORCE V MAKES EQUILIBRIUM OF RESISTING FORCES R_1, AND P_1,

6. V_r: VERTICAL RESISTING FORCE V_r MAKES EQUILIBRIUM OF ACTIVE FORCES R_1^i, AND P_1^i

Figure 2.7 These figures show internal forces at any section of a beam as they are explained by Seely and Smith,[25] Kommers[37] and Timoshenko.[20]

elements. This law is illustrated in Figure 3.2. Here the term "flexural bending" is used because this is different from "pure bending" where transverse forces do not appear.

In the *Plan Review Manual*, edited by the International Conference of Building Officials, USA, 1977, the action of tensile forces from the left section to the left, and from the right section to the right are very clearly illustrated (Figure 17.6). Their action is shown on the free body (Figure 17.6(d)), where the tensile force T is acting perpendicularly toward the given section, while compression force C is acting perpendicularly from the given cross section. It is shown, as well, in the stress diagram that the active compression force C is acting from the triangular diagram outwards, while the tensile force T is acting towards the triangle (Figure 17.6(c)), just as it should be.

The natural orientation of the internal active forces of compression and tension in a bent member starts appearing (in the newer publications) in continuous beams as well, but without a clearly illustrated understanding of the difference between the active and resisting forces inside the beam itself – even though the active forces are oriented contrary to the resisting ones. Still less is mentioned of their real contribution to the diagonal failure of the bent concrete element. In the manual *Masonry Designer's Guide* 1993, ACI, the directions of actions of internal active compression and tensile forces in the continuous beam are shown very clearly. The compression force is acting from the section of the beam (toward the critical section of the beam left and right), and the tensile force is acting towards the section of the beam (from the critical section left and right toward the supports) for each simple beam of the continuous beam. This is illustrated very clearly in the *Masonry Designer's Guide* in Figures 8.4-16, 8.4-20 and 8.4-21. But (as a result of theory which did not recognize the existence of internal active and internal resisting forces in a flexurally bent member), the same author (or any other author in the world) will use, for showing equilibrium using a free body diagram, compression and tensile forces oppositely oriented to the above (real active) forces and call them active forces, instead of resisting forces. Also, the same forces will be used to determine the stress condition of such a beam. In other words, this means that today we determine resisting stresses but we call them active stresses of any flexurally bent member.

It is useful here to emphasize that in the preparation of the *Masonry Designer's Guide*, there were numerous collaborations of Committees, twenty-five independent senior authors who represented a brain trust in their field in the United States. In addition, there were nineteen experts who served as reviewers, thus making their own scientific contribution to the manual. However, no one suggested evidence for contradictions in the direction of actions of such forces by comparing the above figures with Figure 11.3-6 of the same manual.

From the above comments, it follows that both types of forces (active and resisting) are simply classified as active in flexurally bent members, even though the second type serves as balancing forces for a given element. This is another reason why this book has been written.

As stated previously, our theory is based on a very simple premise; that in every flexurally bent member, internal active and internal resisting forces exist simultaneously (see original Figure 3.2 in Section 3.4.2.1 of Chapter 3).

2.4.1 General concept of equilibrium of a free body

Concerning shear forces and resisting shear forces in a flexural member, two fundamentally different concepts of equilibrium exist:

1   equilibrium of a free body of the entire structural member (as shown in Figures 2.5 and 2.7(a)); and
2   equilibrium of a free body of only a portion of a given flexural member (as shown in Figure 2.7(b)).

2.4.1.1 Equilibrium of a free body of a member as a whole

Equilibrium of a free body as a whole is achieved by force vectors placed at the points on a line to represent the external forces (including supports) which act on a given member. This equilibrium is governed by Newton's first law where the movement of a body is prevented: when a body is at rest, the resultant of all of the forces exerted on the body is zero. This is the exact meaning of Newton's first law. In such a case, vertical shear forces V_1 (caused by the support and oriented as the support) are in balance with vertical resisting shear forces V_r (caused by the resistance of the material and oriented downwardly), as illustrated in Figure 2.7(b). Consequently, vertical shear forces mentioned by Timoshenko "... the tendency of the left portion to move upwardly relative to the right portion ..."[20] and by Seely–Smith "... the vertical shear V for section A-A is an upward force equal to 1500 lb ..."[25] are vertical vectors, V_1 acting on the left portion of the member upwardly and V_2 acting on the right portion of the member downwardly or vice versa. One such vector is illustrated in Figure 2.7(b) by a dashed line as force V and is shown there only to prove that such active force V indeed exists in an uncut beam exactly as Seely–Smith and Timoshenko have described it.

By some error rather than by Newton's third law, supports for a flexural member are called "reactions." These have nothing in common with the active force and its reaction, as resistance of material to deformation. This unfortunate definition of reaction (instead of physical support) also contributed to hiding the fallacy of diagonal tension for almost a century; "reaction" is an active force, oriented upwardly (R_1 and R_2) with real reaction (R_{1r} and R_{2r}) oriented downwardly (as illustrated in Figure 2.7(a)). At the same time, the free body is in equilibrium by reactions of C_r and T_r as illustrated in Figure 2.6. Consequently, principal stresses (forces C and T) as shown in textbooks are reactions (resisting forces) so they cannot contribute anything to originate the so-called "pure shear" which could eventually lead to diagonal tension.

Also, probably as a result of the definition of reactions at supports, pioneers named the real reactions (C_r and T_r) for equilibrium of a free body as "active forces (C and T)" which led them to combine compression resisting force C_r with external force F,[10,18] and tensile resisting force T_r with shear force V to create (non-existing) compression struts[18] in order to prove the truss analogy theory (see Figure 2.3).

2.4.1.2 Equilibrium of the free body as a part of the same member

A GENERAL REVIEW

Concerning the equilibrium of a portion of a free body, the following two comments are important:

(1) The existing structural theory that shear force (V) at the right surface of the left portion is in equilibrium with another shear force (V) at the left surface of the right portion of a free body (acting in an opposite direction)[26] could be valid only if it governed their behavior through gravitational or magnetic forces as an electron and proton do. Yet, both forces must be located in the same plane. Push or pull forces are governed by a very simple rule: any action must have its own physical reaction, located in the same line, with the same magnitude but oppositely oriented. This author knows of no possible scientific evidence to prove the illusion of our pioneers that equilibrium is possible for two forces, located in two planes, irrespective of how close or how far they are located from each other. There is also no difference between internal or external forces (if it would be an excluded concept that internal force could be reaction to external force) because "the term force as used in mechanics, refers to what is known in everyday language as a push or pull."[24] Thus, any action must have its own reaction irrespective of where such force is located: movement of such force must be stopped by a counteracting force or its reaction. Because gravitational or magnetic attraction does not exist between any two push or pull forces, such a concept that left shear force (V) at the right face is in equilibrium with right shear force (V) at the left face of the cross section is a simple illusion in the same way as diagonal tension itself. Besides, if such equilibrium cannot be proved algebraically, it must be nonexistent.

(2) By existing structural theory, the forces shown as C and T, acting perpendicularly at a given cross section of any free body, cannot be in equilibrium even though the compression force C is equal to tensile force T and the sum of these two forces is zero. Their internal moment M_r is not their reaction but these forces could be replaced by their moment M_r. Such moment M_r is in equilibrium with its own external moment M at a given cross section. So, as noted previously, the forces which make an external moment M are only capable of making an equilibrium of shown (resisting) compression and tensile forces acting at any cross section. Active forces (C and T) could be in equilibrium only by their corresponding resisting forces (C_r and T_r) and the free body must be in equilibrium

(Figures 2.6 and 2.8(b)). Evidently, the existing concept of the equilibrium of a free body, where horizontal forces are concerned, is also algebraically unprovable: translational equilibrium is achieved but not rotational equilibrium.

To conclude, it can be said that the internal resisting moment M_r, created by the reaction $R_{1r} \cdot x$ is equal to the internal resisting moment M_r, created by the internal resisting forces $C_r \cdot d$ or $T_r \cdot d$. But the balancing moment for the resisting moment $R_{1r} \cdot x$ is the internal moment created by the internal active forces, $C \cdot d$ or $T \cdot d$.

B PROOF THAT A PART CUT OUT FROM A BENT BEAM IS NOT EQUILIBRATED BY THE FORCES OF AN OPPOSITE PART CUT OUT

Given that all shear forces from a bent beam are in translational but not in rotational equilibrium, different attempts were made to solve this problem using the existing knowledge. One attempt was to equilibrate the forces on the left part with the forces on the right part, and vice versa. In fact, this is an error, as it becomes an obstacle to the study of a bent beam. This error is clearly demonstrated at the equilibration of the vertical shear forces on the part cut out from the bent beam. This made the additional discussion, concerning our study and the mentioned problem, necessary.

Some authors define the internal active shear forces as external shear forces even though they have shown that they were located inside of the beam on the very section, as shown in Figure 2.18 by Winter–Nilson.[23] But, in defining them as external forces, the tendency to emphasize that these were real forces causing the shear of the beam in each cross section was obvious, since this is the magnitude of the difference between the external bending forces and the support force (reaction), where $V_{ext} = R_1 - P$ (Figure 2.18). They named the internal resisting shear forces, "internal shear forces," equilibrating directly (on the very section) the so-called "vertical external shear forces," where $V_{int} = V_{ext}$. Further, they say: "shear is resisted by the uncracked portion of concrete at the head of the diagonal crack, or $V_c = V_{ext}$," or, let us repeat, $V_{int} = V_{ext}$. Here, there is no doubt that internal resisting shear forces resist active shear forces. So it follows that the left part of the beam, in relation to the vertical shear forces, is equilibrating itself, by its internal shear forces V_c or V_{int} that are equilibrated by the external shear forces V_{ext}. This means that the left part does not equilibrate the right part and the following quote clearly confirms how it is sometimes applied in literature: "There is an external upward shear force V_{ext}, acting on this portion, which, for the particular loading shown, happens to be $V_{ext} = R_1 - P_1$."[21]

Further, the statement concerning the equilibrium of the active shear force V on the left portion with its internal resisting shear force V_r: "equilibrium requires $V_{int} = V_{ext}$"[23] (that is, $V_r = V$), is a proof that the internal vertical shear force is really equilibrated by its internal resisting shear force, and not that the left portion equilibrates the right portion. This means further that vertical shear forces are in translational equilibrium, since they are equal and lie in the same plane. They are in rotational equilibrium as well since they lie on the same line and are opposite in orientation. Thus, both requirements of Newton's first law are satisfied since the body is really in translational and rotational equilibrium.

The point of view of these authors (Winter–Nilson) is in full conformity with the points of view of Seely–Smith and Timoshenko; namely, that the left portion has never equilibrated the right portion by its forces, and vice versa, so that each portion is equilibrating itself by its own forces. In this sense, Seely–Smith say very clearly: "The symbol V will be used to denote vertical shear forces and, for convenience, the forces that lie on the left section will here be used."[25] So it is clear that they are speaking only about the left section and not about the right section as well. Timoshenko also clearly explains that the left portion equilibrates itself: "But this link and roller connection alone will not counteract the tendency of the *left portion* to move upward.... To prevent this relative sliding, some additional device will be needed and the bar will exert a downward force V on the left portion."[20] There is no doubt that Timoshenko is speaking exclusively on equilibrium of the *left portion*. Thus, the assertion of any author or designer, that the left portion of the free body, cut out from the bent beam, is equilibrating the right portion, is untenable.

2.4.2 Existence of internal active and internal resisting forces – mathematical proof

The branch of physics that treats action of forces on material bodies and their motion, including statics, kinetics and kinematics, is generally defined as mechanics. The science of mechanics is based on three natural laws, stated for the first time by Newton in his 1686 manuscript, "The Mathematical Principles of Natural Science." But in his laws, specifically his third law, which deals with action and reaction, Newton did not mention the equilibrium of a free body of a flexural member, which differs fundamentally from the concept of push–pull forces and their equilibrium. Because of this, no one has succeeded in establishing the real equilibrium of a portion of a flexural member. As a result, today we have a situation where the existing theorem of equilibrium cannot be proved mathematically. In other words, to establish the equilibrium of a body, the sum of all forces, acting in any possible direction, must be zero. The moment of all forces about any axis must also be zero. Algebraically, $\sum X = 0, \sum Y = 0$ and $\sum M = 0$. But, "these equations $\left(\sum X = 0, \sum Y = 0 \text{ and } \sum M = 0 \right)$ cannot be proved algebraically; they are merely statements of Sir Isaac Newton's observation that for every action on a body at rest there is an equal and opposite reaction."[26] This is illustrated by Figure 2.7(b), where horizontal or vertical forces are not in equilibrium because the internal active and resisting forces in a flexural member have not been recognized. This has been the main motivation for the author to spend the last thirty years in research and experimentation and which led him to discover a new law of physics. As has been said, this law could be described thus: *in a flexural member, nature simultaneously creates internal active and internal resisting forces of compression, tension, horizontal shear and vertical shear, where any couple of forces is located in the same line, oppositely oriented and of equal magnitude.*

The ramifications of this discovery are so deep that it could be said that the entire existing structural theory will be changed by the proof that diagonal tension (as shown in any textbook) does not exist and that the truss analogy is, fundamentally, an incorrect concept.

2.4.2.1 Proof I

The derivation of equations for internal and external moments in a flexural member could be used to prove the existence of internal active and internal resisting forces. In that sense, it could be said that in flexural bending a very interesting phenomenon appears for the first time, namely that the internal resisting moment M_r and the external active moment M, for a given cross section, could each be determined by two different methods.

A CASE OF THE EXTERNAL ACTIVE MOMENT M

The external active moment M, for a given cross section, can be determined from the external vertical forces by the classical method (Figure 2.6):

$$M = R_1 \cdot x. \tag{2.1}$$

Also, the external moment M, for the same cross section, could be determined from the internal active compression (C) and tensile (T) forces (Figure 2.6):

$$M = C \cdot d = T \cdot d. \tag{2.2}$$

The final equation can be written as

$$M = R_1 \cdot x = C \cdot d = T \cdot d. \tag{2.3}$$

B CASE OF THE INTERNAL RESISTING MOMENT M_r

The internal resisting moment M_r, for a given cross section, could be determined by the classical idea from internal resisting forces (Figure 2.6):

$$M_r = C_r \cdot d = T_r \cdot d. \tag{2.4}$$

The internal resisting moment M_r, for the same cross section, could also be determined from the vertical internal resisting forces (Figure 2.6):

$$M_r = R_{1r} \cdot x. \tag{2.5}$$

The final equation can be written as

$$M_r = C_r \cdot d = T_r \cdot d = R_{1r} \cdot x. \tag{2.6}$$

If one can accurately determine the external moment M for a given cross section by applying internal active compression (C) and tensile (T) forces, it is clear that such internal active forces must exist. If so, the determined moment is identical to the moment determined by the external vertical forces, and this must mean that direct mutual correlation does exist between such forces.

Also, if one can accurately determine the internal resisting moment M_r by applying vertical resisting forces (the existence of which is evident by Newton's third law) and such moment is identical to the moment determined by internal compression and tensile forces, it is clear evidence that such forces must be internal resisting compression (C_r) and tensile (T_r) forces. The internal resisting moment M_r cannot be a resisting moment determined, at one instance, by resisting vertical forces and at the next instance, by internal horizontal active compression and tensile forces. In other words, if $A - B = C - B$, then A must be equal to C. Yet as a result, internal moment M_r, determined in either way, is identical in value and thus it follows that such forces do exist, as well as the direct mutual correlation between such forces.

2.4.2.2 Proof II

By applying the same internal active and internal resisting forces, it becomes possible to algebraically prove that $\sum X = 0$, $\sum Y = 0$ and $\sum M = 0$:

$$\sum X = 0: \quad H = H_r \qquad \text{(Figure 2.7(c))}, \qquad (2.7)$$
$$C = C_r \qquad \text{(Figure 2.6)}, \qquad (2.8)$$
$$T = T_r \qquad \text{(Figure 2.6)}. \qquad (2.9)$$
$$\sum Y = 0: \quad R_1 = R_{1r} \qquad \text{(Figure 2.6)}, \qquad (2.10)$$
$$V = V_r \qquad \text{(Figure 2.6)}, \qquad (2.11)$$
$$P_1 = P_{1r} \qquad \text{(Figure 2.7(a))}. \qquad (2.12)$$
$$\sum M = 0: \quad M = M_r \qquad \text{(Figure 2.6)}, \qquad (2.13)$$
$$M = R_1 \cdot x = C \cdot d = T \cdot d \qquad \text{(Figure 2.6)}, \qquad (2.14)$$
$$M_r = R_{1r} \cdot x = C_r \cdot d = T_r \cdot d \qquad \text{(Figure 2.6)}. \qquad (2.15)$$

Since internal active and internal resisting forces are the functions of the shown equations, this is an additional mathematical proof that such forces do indeed exist.

Also, this is the first time that equilibrium of a free body has been proven algebraically. Proving the equilibrium of a free body proves the existence of internal active and internal resisting forces at the same time. Yet, the existence of such internal forces made the explanation of diagonal cracking and diagonal failure of a flexural member possible. This will be the exclusive subject of discussion in Section 2.4.4.

2.4.3   Shear stresses in a flexurally bent member

For equilibrium of a unit element from a flexural member, one does not need to apply Newton's first law to prevent possible rotation because such rotation does not exist on the member itself and because such a unit element is already in equilibrium by the law of action and reaction as illustrated in Figure 2.7(c). Moreover, the direction of active or resisting stresses is determined by the bending phenomenon, as explained by Timoshenko, Saliger and Winter–Nilson. Consequently, such stresses must be observed since they are directed by the nature of bending and not as indicated by Ritter–Morsch[1,3,22]. Yet, the sliding shear stresses of the existing theory could not be proved by observing such sliding (opposite to that in Figures 2.5(a) and (b)) or by measuring their strains directly. Instead, their alleged existence is calculated by trigonometry (Mohr's circle) from the principal strains (elongation of a given unit), which is what the assumption of pure shear existence[1,22] is all about. So, by multiplying such strain with its modulus of elasticity, the principal stress is calculated, from which shear stress is obtained.[20,21] Consequently, it appears that shear stresses, so determined, are oppositely oriented to shear stresses caused by bending itself.

The problem can be illustrated by quoting Timoshenko when he avoided showing any example of direct measurement achieved: "However, accurate measurement of the shearing strain is found to be very difficult. It is easier and *more accurate* to measure the principal strains."[21] ... "The resistance of concrete to pure shearing stress has never been directly determined,"[27] ... "No one has ever been able to accurately determine the resistance of concrete to pure shearing stress."[12]

One could measure shear strain directly on only one block exposed to shear forces (Figures 7.13 and 7.14 of reference 21), but no one ever succeeded in measuring shear strains in a flexural bending member, because they do not exist.

As illustrated in Figures 2.5(b) and 2.7(a), horizontal sliding shear stresses τ_h are *always* present in a bent member (if one excludes pure bending, or a portion of the beam with constant moment): The upper portion of the member has a natural tendency to slide from the lower portion by moving toward the supports while the upper plane of the lower portion has the tendency to move from the support toward the critical cross section. This fact of sliding in opposite directions through the neutral plane is shown in several textbooks. Professor Saliger's Figure 209[19] (presenting sliding stresses τ_h with their direction of sliding) is clearest and is reproduced in Figures 2.5(b) and 2.7(c). For clarity, a unit element taken from Figure 2.7(a) is shown in Figure 2.7(c) with the same horizontal shear stresses (τ_h) and vertical shear stresses (τ_v) as shown in Saliger's Figure 209[19]. Besides, such sliding shear stresses (τ_h and τ_v) are supported by any shear slip model[28] and by common sense. Thus, the active shear stresses (τ_h and τ_v) are analogous to similar active shear stresses (τ_h and τ_v) in a twisted shaft and such stresses and their causes should be studied with a view to possible prediction and prevention of diagonal cracking in a flexural member.

Such rigid logic leads one to see that diagonal tension is capable of the elongation of one of the diagonals of a unit element, cut from a bent member. This is presented in Figure 2.7(c). Yet, such tension-stretching of one diagonal of a unit element is in a diagonal which is opposite to the one in the classical concept.

Immediately the question arises: what is wrong with the classical concept of equilibrium of a unit element taken from a bent member? The answer is: there is nothing wrong. A unit element is in real equilibrium by resisting shear stresses τ_{hr} and τ_{vr} as shown in Figures 2.7(a) and (c). The only error is that such stresses (τ_{hr} and τ_{vr}) cannot cause any deformation, stretching, diagonal cracking or diagonal tension of a given unit. They can only cause equilibrium. Because of its utmost importance, it should be reiterated that by using the classical concept for equilibrium of a unit element from a twisted shaft, active (real) shear stresses (τ_h and τ_v) are applied. While, for equilibrium of a unit element from a bent member, resisting shear stresses caused by bending (τ_{hr} and τ_{vr}) are applied, as illustrated in Figures 2.7(a) and (c).

Let us explore the equilibrium of a unit element through the classical point of view by applying stresses caused by the bending phenomenon. For that purpose only Timoshenko[20] and Winter–Nilson[23] will be quoted.

Timoshenko defined the horizontal resisting shear stresses (τ_{hr}), as "horizontal shear stresses to prevent sliding" and illustrated them in his Figure 159.[20] The direction of action of such stresses, as shown by Timoshenko, is reproduced in Figures 2.5(a) (unit element), 2.7(a) and (c) as τ_{hr}.

The same horizontal resisting shear stresses τ_{hr} are defined by Winter–Nilson as "horizontal shear stresses which prevent shearing or sliding," and they illustrated these stresses in their Figure 2.13.[23] The directions of action of such stresses are also reproduced in Figures 2.5(f), 2.7(a) and (c) as τ_{hr}.

Indeed, a unit element is in real equilibrium concerning resisting shear stresses (τ_{hr}) as the classical theory explains such equilibrium of a unit element. Such stresses which *prevent sliding* can never cause stretching of a diagonal. However, active shear stresses (τ_h and τ_v) can, as shown in Figure 2.7(c).

Naturally, the total sum of shear stresses at a given plane (or surface) will represent the shear force acting at such a plane or on the surface of a unit element, as shown in Figure 2.7(c) (horizontal shear forces H and vertical shear forces V).

The sum of resisting shear stresses at a given plane (or cross section) will give the resisting shear force acting at a given surface of a unit element, as illustrated in Figure 2.7(c) (horizontal resisting shear force H_r and vertical resisting shear force V_r).

As can be seen from the works of Timoshenko, Saliger and Winter–Nilson, the existence and direction of sliding horizontal and vertical shear stresses (τ_h and τ_v) and horizontal and vertical resisting shear stresses (τ_{hr} and τ_{vr}) in a bent member are very well known facts. Thus, what we have done is put together pieces which are dispersed throughout engineering mechanics literature and pointed out the fallacy of the classical concept of diagonal tension.

2.4.4   An application of the new law of physics: the diagonal failure of a flexural member

As illustrated in Figures 2.5(c), 2.8(a)–(c), diagonal cracking in a flexurally bent member is caused by

1   the tendency of the support to move its portion of the beam upwardly, causing eventual punch shear cracking as illustrated in Figures 2.8(a) and (b); or by
2   the tendency of the external load (uniform or concentrated) to move its portion of the beam downwardly, eventually causing its own punch shear cracking, as illustrated in Figures 2.8(b) and (c); or by
3   the combined tendency of both (support and external load) causing a single diagonal crack, connecting the support and the external load, as shown in Figure 2.8(b).

In other words, diagonal cracking in a bent member is caused by the resultant punch shear force V_n, acting perpendicularly to any diagonal crack.[5] Forces such as V_n are caused (formed) by a combination of the following forces:

1   vertical shear force V_1 caused by the support and oriented as the support itself; here, upwardly;
2   vertical shear force V_2 caused by the external load and oriented as the load itself; here, downwardly;
3   flexural tensile force T caused by flexural bending, as shown in Figures 2.5(c) and 2.8(c).

Obviously, the resultant punch shear forces V_n will be created for any type of load; concentrated, uniform, combined, or any other type of load. As long as pure bending does not exist (where transverse forces are eliminated), the resultant punch shear forces V_n will always be present as a result of the above-mentioned combination.

Simultaneously with the action of punch shear forces V_n, sliding punch shear forces V_n' exist, acting parallel to diagonal cracking, one upwardly and the other downwardly. These forces are caused by a combination of vertical shear forces V_1' and V_2' with compression forces C as shown in Figure 2.8(c). This fact was published as early as 1978 with the denomination of sliding punch shear forces as A and B.[5,13,16] Consequently, observing and balancing all data existing in technical literature, the fact is that another very logical explanation of diagonal cracking of a flexural member emerges from this work.

As shown in Figure 2.8(c), in order to control punch shear cracking, four unknowns must be determined: angle of cracking (α), vertical shear force (V_1 or V_2), horizontal tensile force (T) and resultant punch shear force (V_n). If we can determine only two unknowns, namely the angle of cracking and vertical shear force V; then by trigonometry, the resultant V_n and flexural tensile force T

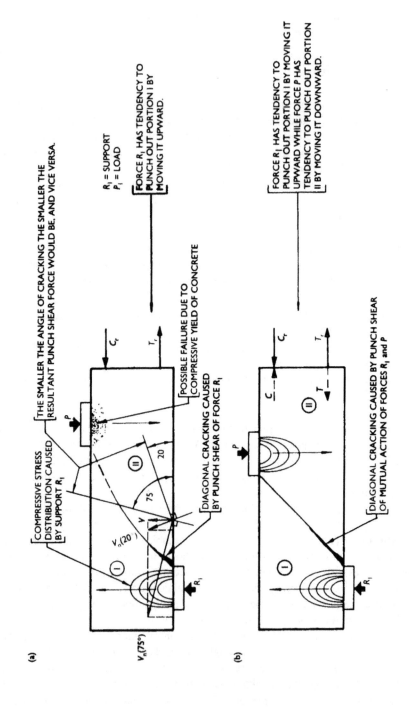

(a)

(b)

THE SMALLER THE ANGLE OF CRACKING THE SMALLER THE
RESULTANT PUNCH SHEAR FORCE WOULD BE, AND VICE VERSA.

R_i = SUPPORT
P_i = LOAD

FORCE R_i HAS TENDENCY TO
PUNCH OUT PORTION I BY
MOVING IT UPWARD.

COMPRESSIVE STRESS
DISTRIBUTION CAUSED
BY SUPPORT R_i

POSSIBLE FAILURE DUE TO
COMPRESSIVE YIELD OF CONCRETE

DIAGONAL CRACKING CAUSED
BY PUNCH SHEAR OF FORCE R_i

$V_n(75°)$

$V_n(20°)$

C_r

T_r

I

II

75

20

P

R_i

FORCE R_i HAS TENDENCY TO
PUNCH OUT PORTION I BY MOVING IT
UPWARD WHILE FORCE P HAS
TENDENCY TO PUNCH OUT PORTION
II BY MOVING IT DOWNWARD.

DIAGONAL CRACKING CAUSED BY PUNCH SHEAR
OF MUTUAL ACTION OF FORCES R_i and P

C_r

T_r

C

T

I

II

P

R_i

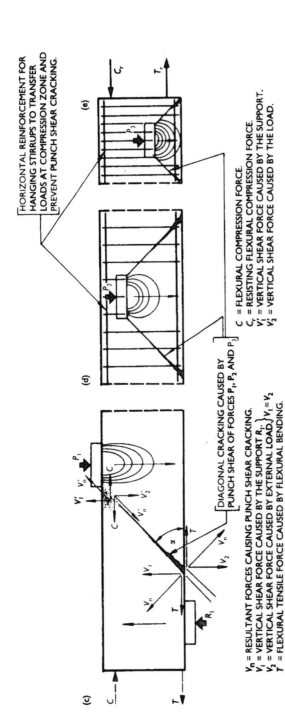

(e)

HORIZONTAL REINFORCEMENT FOR HANGING STIRRUPS TO TRANSFER LOADS AT COMPRESSION ZONE AND PREVENT PUNCH SHEAR CRACKING.

(d)

(c)

V_n = RESULTANT FORCES CAUSING PUNCH SHEAR CRACKING.
V'_1 = VERTICAL SHEAR FORCE CAUSED BY THE SUPPORT R_1.
V'_2 = VERTICAL SHEAR FORCE CAUSED BY EXTERNAL LOAD. $V_1 = V_2$
T = FLEXURAL TENSILE FORCE CAUSED BY FLEXURAL BENDING.
α = ANGLE OF CRACKING.

DIAGONAL CRACKING CAUSED BY PUNCH SHEAR OF FORCES P_1, P_2 AND P_3

C = FLEXURAL COMPRESSION FORCE.
C_r = RESISTING FLEXURAL COMPRESSION FORCE.
V'_1 = VERTICAL SHEAR FORCE CAUSED BY THE SUPPORT.
V'_2 = VERTICAL SHEAR FORCE CAUSED BY THE LOAD.

NOTE:

THE FLEXURAL TENSILE STRESS FIELD WHICH EXISTS CONCURRENTLY WITH THE FLEXURAL COMPRESSION FIELD, ASSOCIATES WITH THE COMPRESSIVE STRESS FIELD CAUSED BY CONCENTRATED FORCES ON THE BEAM, HAVE BEEN SHOWN IN FIG. B OF REF 16.

Figure 2.8 This figure shows three main types of punch shear which Ritter–Morsch did not visualize. The laws of physics show that punch shear can be caused by physical support R_1 (a), by mutual action of support R_1 and exterior load P (b), and by any external load P (c). It is more reasonable to assume that diagonal cracking is caused by external forces intending to punch out some portion of the beam, rather than by the diagonal tension whose existence has never been proved.

can be determined. The vertical shear force V, representing shear forces V_1 and V_2 for a given cross section (for uniform or concentrated load), could be determined by the classical shear diagram. The direction of such shear force V can be determined using our concept that any shear force caused by the support must have the direction as the support itself; and any shear force caused by the external (static or dynamic) load must have the direction of the load itself. Yet, the fact that shear forces must follow the direction of external forces should be a well-known rule which serves as a basis on which the entire theory of earthquake engineering is founded. Figures 7.6, 7.7 and Table 7.1 of reference 29 clearly demonstrate this phenomenon.

Evidently, the basic problem is the angle of cracking, which should be solved first. Taking into consideration the existing knowledge of the mechanism of stress distribution of a concentrated load at the base of the load,[30] it is possible to predict, with a high degree of accuracy, the location and direction of cracking as illustrated in Figure 2.9. A pullout test from a concrete member is the best example of this 45° stress distribution of a concentrated load.

Three types of punch shear cracking should be distinguished:

a   that caused by a concentrated load;
b   that caused by a uniform load; and
c   the one that occurs in a deep beam.

2.4.4.1 Punch shear cracking caused by a concentrated load

Two possibilities can also be distinguished here:

a   where the concentrated load is nearer to the support than the depth of the member: $x' < d$, as shown in Figure 2.9(a); and
b   where the concentrated load is located on the beam with $x > i + j$, as shown in Figure 2.9(a).

A CONCENTRATED LOAD IS LOCATED IN SECTION "i" ($x' < d$)

As long as the compressive stress distribution (caused by the support and by the external concentrated load) influence each other directly, the cracking line would be a straight line between these two forces. So, by knowing the location of the external concentrated load, angle of possible cracking, β, could be easily calculated (or measured with a protractor).

Knowing the vertical shear force V_1 (or V_2) for a given cross section and the angle of probable cracking, β, the resultant punch shear force V_n can be calculated:

$$V_n = \frac{V_1}{\cos \beta}. \tag{2.16}$$

To secure ductile failure in seismically sensitive zones, transverse reinforcement will be calculated in such a way that the entire force would be controlled by reinforcement:

$$V_n = \text{fs} \cdot \text{As} \tag{2.17}$$

"As" is area of steel, while "fs" is stress in the steel.

Because the slope of cracking in this case increases rapidly as the concentrated load moves toward the support, this area should be covered by reinforcement perpendicular to the lines of possible cracking. Note that cracking in a deep beam can be almost vertical;[17] consequently, vertical reinforcement becomes useless here.

When the concentrated load is located closer to the support than approximately 75 , then cracking will be governed by the support (at 45) rather than by a straight line between these two forces. This is a result of the creation of an arch by an external load. The resulting fan of stress distribution lines cannot be united to make a possible straight line of cracking between these two forces.

In seismically insensitive zones, a portion of the vertical shear force V can be controlled by concrete. But because punch shear can never be developed in a compressed zone, it follows that, in a flexural member, punch shear is maximal at the edges of the external tensile fibers and zero at the neutral plane. So we can allow only the flexural tensile zone of the member to take over some portion of the vertical shear V for a given cross section:

$$V' = abv_c, \tag{2.18}$$

where V' is the portion of the vertical shear force V controlled by concrete, b is the width of the flexural rectangular member, a the distance of the neutral plane to the extreme tensile fibers and v_c the allowable punch shear stresses of concrete.

It should be emphasized here that punch shear strength is an extremely variable parameter: for the same quality of concrete, the punch shear test performed by the pullout test for an unloaded member, is one value. However, the same pullout test, taken from the flexurally bent member, is a completely different value. A rule could be: the larger the flexural tension of the member, the smaller the resistance to punch (or any other) shear. For this reason, it is impossible to ever depend on some strength of concrete against punch shear in a flexural member. Thus, a punch shear test must be performed for the worst flexural bending, in which case the real punch shear value would drop down to probably one-tenth or so of that found by the pullout test for an unloaded member. For example, if 6,000 psi concrete is loaded up to 5,000 psi, only 1,000 psi remains to resist the pullout test (or punch shear cracking) in the tensile zone.

It should be noted here that we have denominated diagonal cracking in a bent member as punch shear cracking, because external forces act to punch out some portion of the beam in just the same way that punch shear occurs in a pullout test with a cone formed at a 45 slope.

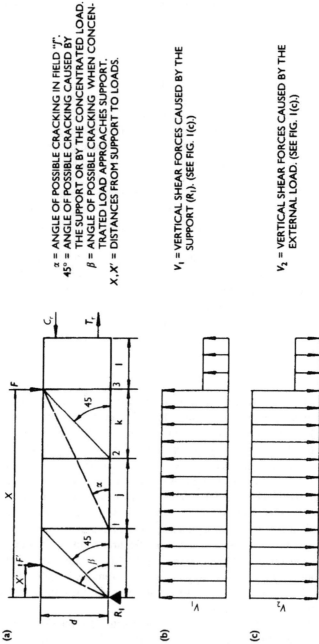

α = ANGLE OF POSSIBLE CRACKING IN FIELD "J".
45° = ANGLE OF POSSIBLE CRACKING CAUSED BY THE SUPPORT OR BY THE CONCENTRATED LOAD.
β = ANGLE OF POSSIBLE CRACKING WHEN CONCENTRATED LOAD APPROACHES SUPPORT.
X, X' = DISTANCES FROM SUPPORT TO LOADS.

V_1 = VERTICAL SHEAR FORCES CAUSED BY THE SUPPORT (R_1). (SEE FIG. I(c).)

V_2 = VERTICAL SHEAR FORCES CAUSED BY THE EXTERNAL LOAD. (SEE FIG. I(c).)

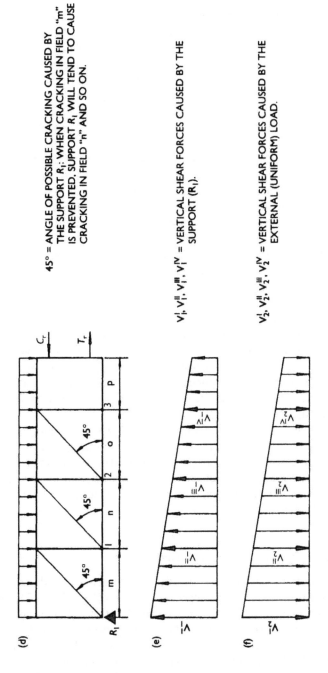

45° = ANGLE OF POSSIBLE CRACKING CAUSED BY THE SUPPORT R_1; WHEN CRACKING IN FIELD "m" IS PREVENTED, SUPPORT R_1 WILL TEND TO CAUSE CRACKING IN FIELD "n" AND SO ON.

$V_1^I, V_1^{II}, V_1^{III}, V_1^{IV}$ = VERTICAL SHEAR FORCES CAUSED BY THE SUPPORT (R_1).

$V_2^I, V_2^{II}, V_2^{III}, V_2^{IV}$ = VERTICAL SHEAR FORCES CAUSED BY THE EXTERNAL (UNIFORM) LOAD.

Figure 2.9 This figure illustrates the most probable angles of cracking for flexural member exposed to concentrated or uniform load. (a)–(c) Correspond to diagonal cracking and concentrated load. (d)–(f) Correspond to diagonal cracking and uniform load.

B   CONCENTRATED LOAD IS LOCATED IN SECTION "k" ($x > i + j$)

Because the stress distribution of a concentrated load at its base is at approximately 45°, it is clear that for the entire field (shown as "j" in Figure 2.9(a)), it should be assumed that the cracking line would be a straight line from field "i" (Point 1) to the exterior load (because of web reinforcement in field "i", cracking is forced to occur in the next field denoted as "j" and also because "failure occurred with a sudden extension of the crack toward the point of loading").[6] This is exactly as practical experiments have already proved, that "in most laboratory tests, if dead load is neglected, the shear span is the distance from a simple support to the closest concentrated load."[4]

For field "k" vertical reinforcement will be determined for an angle of 45 because this is the most probable punch shear cracking caused by the external load F: "Under a single point load, diagonal failure occurs closer to the applied load rather than to the support (Figure 1(a))."[31] Even a concentrated load at mid-span could cause pure punch shear cracking at 45° (because this force is 50 percent larger than at either support), as Figure 5 of reference 32 illustrates.

Field "j" can be covered with minimal web reinforcement to avoid possible buckling and provide greater rigidity of a beam if twisting occurs.

In conclusion, it could be said that the same reinforcement will be required in the field of support and in the field of the external concentrated load, while between these two fields a much smaller amount of web reinforcement will be required (because of a lack of any other forces – loads between R_1 and F). This is in spite of the fact that, in a shear diagram, the transverse shear force does not change between the support and the first external load (see Figures 2.9(a) and (e)). Furthermore, the shear diagram and punch shear have a discrepancy: maximal reinforcement is a function of punch shear stresses and not of the shear forces shown by the shear diagram.

In Figures 2.8(d) and (e), vertical stirrups are shown which prevent possible punch shear cracking and also transfer forces P_2 and P_3 at the top of the beam; so they act in a compressed zone by being hung from the horizontal compression reinforcement. In other words, by transferring forces P_2 and P_3 to act in the compression zone instead of the tension zone, the resistance to punch shear cracking is increased considerably.

2.4.4.2   Punch shear cracking caused by a uniform load

The concept of a uniform load is based on only one concentrated load, namely on the concentrated load caused by the support. So for field "m" (Figure 2.9(d)), vertical web reinforcement will be determined on the basis that probable cracking is at 45° from the support. It is also true that if dynamic (sudden) uniform loading occurs, which is capable of causing failure, then punch shear will be vertical, as shown in reference 7. Thus, if this can happen then web reinforcement must be on a slope to prevent failure due to such action. Naturally, shear force V would be equal to reaction R_1 (largest). By installing such reinforcement in field "m", possible cracking would be moved to the next field denoted by "n" where the maximum vertical shear force would be V_1'' (see Figure 2.9(e)) and also the crack will be at approximately 45 . The reason is that now the support and field "m" represent a rigid unit, so the tendency of this rigid unit is to punch out an adjacent portion of the uniformly loaded beam and move it vertically. Yet, for any new field, reinforcement would be decreasingly smaller following the magnitude of shear forces at its cross section, as shown in Figures 2.9(d) and (e).

To reiterate, a uniform load punch shear cracking, in general, would be at a 45 angle before starting to combine with flexural cracking in the vicinity of the maximum moment where it follows an irregular line, not necessarily at a 45° angle.

2.4.4.3   Punch shear cracking in a deep beam ($h/L \geq \frac{1}{2}$)

For a deep beam, the slope of punch shear cracking is a straight line between the support and the external load and is usually much larger than 45 . Thus, to calculate punch shear force V_n, such a slope should be used. Web reinforcement would be perpendicular to such a slope because, if it is placed vertically, punch shear cracking would develop vertically – parallel to the first stirrup adjacent to the support, or the first stirrup adjacent to the concentrated load.[17] This is because a very large punch shear force V_n, by increasing the angle of cracking, becomes almost horizontally oriented. When the resultant punch shear force is horizontal, the cracking is vertical (90°) but the shear force V is zero. Such a fact is illustrated in Figure 2.8(a).

Evidently, for a uniformly loaded deep beam, the angle of diagonal cracking (leading to possible failure) cannot be flatter than the angle covered by the straight line connecting the support and the mid-point of the member at the top. In other words, the slope of diagonal cracking cannot pass to the other side of the maximum moment.

For a uniform load, the smallest angle of possible diagonal cracking would be a straight line between the support and the midpoint of the member at the top; and such an angle is usually much larger than 45 . But, actually, the angle would be a straight line from the support to the top of the member at approximately one-third length from the support. The reason for this is that, as a result of arch action in a deep beam, the arch will prevent possible cracking in a straight line from the support to the midpoint. The arch would force the crack to go to about the one-third point at the top of the member.

It should be mentioned that diagonal cracking in a deep thin web double T beam could start near the neutral axis in the tensile portion of the web because the tensile flange can control punch shear forces V_n while the thin web cannot. Therefore, diagonal cracking would originate there.

If a vertical shear force were oriented, as the classical theory suggested,[33] opposite to the direction shown in Figure 2.8(c), then diagonal cracking could never develop in a flexurally bent member, as is clearly seen in Figure 2.8(c). In other words, if we applied vertical shear forces V_1 and V_2 on a unit element which is incorporated at the left side of the vertical shear forces V_1 (or at the right side of force V_2), the alleged diagonal tension appears to be parallel to the diagonal cracking (Figure 4.1).

2.5   Comments about our shear diagram based on the new law of physics

Because our shear diagrams and the direction of action of vertical shear forces are vital to our theory of diagonal cracking, Figures 2.10 and 2.11 have been introduced in order to show the direction of vertical shear forces V in any flexural member. This is because the shear diagram for a uniform load is the result of the superimposition of two different diagrams (with two different directions of action of the shear forces), which results in the triangular shape with which we are familiar, as illustrated by Figure 2.10.

The shear diagram for vertical shear forces caused by the supports alone is of a rectangular shape, with shear forces oriented upwardly (as shown in Figure 2.10(c)), because what is relevant is the action of a concentrated load on a given member. At first glance, there is a discrepancy in the shape of this diagram with the shear diagram shown in Figure 2.9(e), where the shear diagram for supports is shown to be triangular in shape. To avoid confusion, an evolutionary development of a shear diagram for a uniform load is introduced, as illustrated step by step in Figure 2.10. From Figure 2.10(d), it is evident that, for a uniform load alone, the shear diagram is of a triangular shape with zero force at the supports and maximal force at the middle of the span, oriented downwardly. Evidently, the classical shear diagram shows triangles oriented oppositely compared to that illustrated in Figure 2.10(d), without showing the direction of shear forces. To prove the triangular shape of the shear diagram for a uniform load, as shown in Figure 2.10(g), where a cantilever is exposed to both a uniform load acting downwardly and a concentrated load at its end acting upwardly, we superimpose the shear diagrams for these two conditions (Figures 2.10(d) and (e)) and the final diagram becomes triangular in shape with zero force at the anchorage point and maximal force at the end of the cantilever (Figures 2.10(f) and (g)). A beam on two supports could be imagined to be composed of two cantilevers with an anchoring wall at the middle of the span, where the support of a simple beam becomes the vertical concentrated load acting upwardly while the uniform load acts downwardly (Figure 2.10(e)). By such an observation, it becomes clear that the shear diagram for a uniform load of a simply supported beam becomes triangular in shape by superimposition of two different shear diagrams (as illustrated by Figures 2.10, 2.9(e) and (f)), with the direction of shear forces V, caused by uniform load, oriented downwardly (Figure 2.10(g)).

Following the above guidelines, it becomes clear that a shear diagram for a fixed-end beam will be composed of two different diagrams:

a   a shear diagram for a simply supported beam, located between two PIs; and
b   a shear diagram for two cantilevers supporting a simply supported beam between two PIs; or a shear diagram of two cantilevers with loads concentrated at their ends.

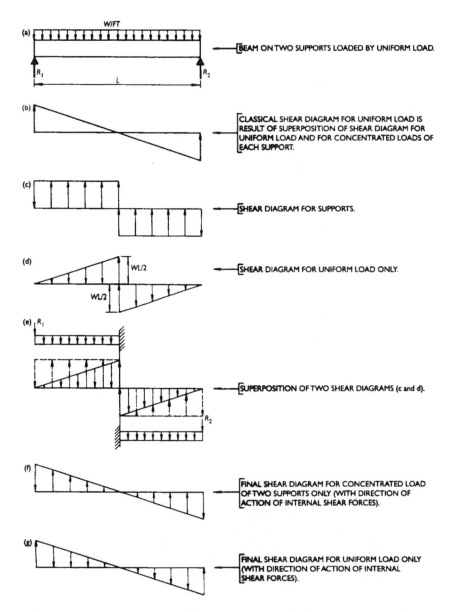

Figure 2.10 Evolution of shear diagram for uniform loads: final shear diagram for uniform load is the result of the superposition of the sheer diagram for the supports (rectangular shape) with the shear diagram for a uniform load (triangular shape). In such a diagram, the direction of shear forces caused by the supports is oriented *upwardly*, while the direction of the shear forces caused by the uniform load is oriented *downwardly*.

BEAM FIXED AT BOTH ENDS – UNIFORMLY DISTRIBUTED LOAD

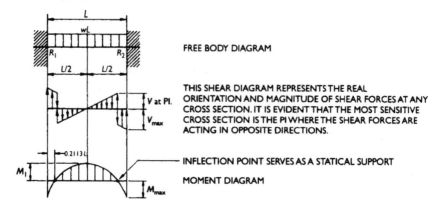

FREE BODY DIAGRAM

THIS SHEAR DIAGRAM REPRESENTS THE REAL ORIENTATION AND MAGNITUDE OF SHEAR FORCES AT ANY CROSS SECTION. IT IS EVIDENT THAT THE MOST SENSITIVE CROSS SECTION IS THE PI WHERE THE SHEAR FORCES ARE ACTING IN OPPOSITE DIRECTIONS.

INFLECTION POINT SERVES AS A STATICAL SUPPORT

MOMENT DIAGRAM

BEAM FIXED AT BOTH ENDS – CONCENTRATED LOAD AT CENTER

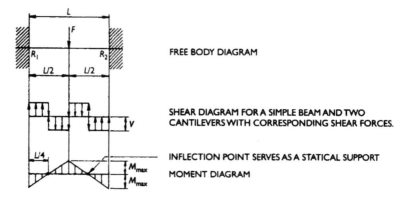

FREE BODY DIAGRAM

SHEAR DIAGRAM FOR A SIMPLE BEAM AND TWO CANTILEVERS WITH CORRESPONDING SHEAR FORCES.

INFLECTION POINT SERVES AS A STATICAL SUPPORT

MOMENT DIAGRAM

Figure 2.11 This figure illustrates that the shear diagram of a beam on two supports and the shear diagram of the fixed beam are two completely different diagrams: The first is composed of only one beam with shear forces acting *downwardly*, while the second one is composed of two cantilevers and one simple beam with shear forces acting *upwardly* in one position and *downwardly* in the other. The simple beam has only one (positive) moment, while the fixed-end beam developed two moments (positive and negative). The first one has no inflection point, while the second has two inflection points.

As has been noted, the direction of shear forces (stresses) must follow the direction of the bending line of the beam: for a cantilever, oriented upwardly; while for a simple beam, oriented downwardly. This is illustrated by Figure 2.11. Yet, without observing the direction of the action of shear stresses (forces) in a stressed beam, it is impossible to predict the safety of a concrete beam, regardless of our scientific background, because real shear exists only at the PI. This has not been recognized in any existing shear diagram in engineering mechanics.

So, to prevent shear failure at the PI, we must connect two beams (a cantilever and a simple beam) together by special stirrups, which extend from the tensile bars of the cantilever to the tensile bars of the simple beam. This becomes the simplest, safest and most economical solution possible to hang up two beams against each other, as is discussed here, in Section 4.5.2.2 of Chapter 4 and elsewhere.[3]

2.6   Reinforcement for the prevention of diagonal failure at bent concrete elements as required by the new law of physics

2.6.1   Deeper beams: theoretical considerations

In the bent beam, left (or right) of the flexural critical cross section, only one point exists in the tensile zone where the internal tensile forces are equal to the internal vertical shear forces. This point is located at a distance from the support equal to about the depth of the beam. Cracks usually develop first at an angle below 45°, as illustrated in Figures 2.12 and 2.13. At each successive point toward the flexural critical cross section, tensile forces become increasingly greater than the vertical shear forces, and their resultant stipulates increasingly steeper diagonal shear failure. In the proximity of the flexural critical cross section, the cracking resultant becomes almost horizontal, and the crack almost vertical. If stirrups follow the directions of these resultants, their effect will be rationally utilized, since the cracking forces will be located along the stirrups. Thus, the beam will be very well protected against cracks (see Figures 2.12 and 2.13).

Even though the cracking resultant V_n increases toward the flexural critical section, it does not determine the real need for stirrups against diagonal failure. *The stirrups absorb only the value of the resultant, where the horizontal and vertical forces are equal in magnitude, and the horizontal reinforcement absorbs the difference of the horizontal tensile forces.* Thus, when diagonal failure occurs, only a portion of the tensile force remains free for tensile reinforcement. The portion of the tensile force engaged in diagonal cracking is equal to the vertical shear force for any given cross section. (In pure bending, diagonal tensile cracking does not exist.) If horizontal reinforcement does not exist in the beam, then the necessary quantity of stirrups will increase in the same measure as the cracking resultant V_n increases for each section.

Point 1, in Figures 2.12 and 2.13, is subjected most to the manifestation of the first crack. This can be explained as follows:

1   The vertical shear force becomes equal in magnitude to the tensile force only here.
2   The cracking angle of 45 is the natural angle, since it corresponds to the critical punch stress penetration of the force acting under the angle of 90° on a plane.

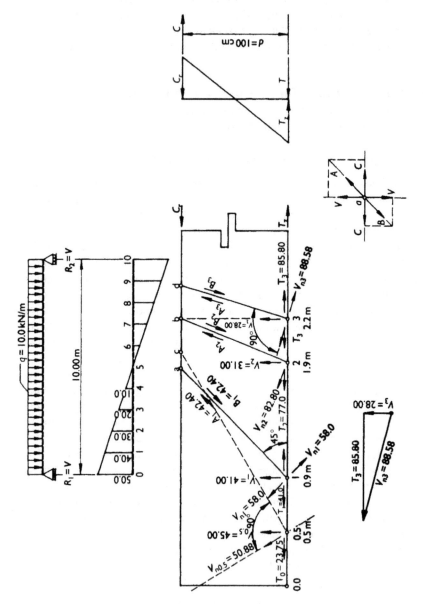

Figure 2.12 This figure illustrates the most critical diagonal failure planes 1–a, 2–b and 3–d for the reinforced concrete beam, loaded by a uniform load. The stirrups must be perpendicular to these planes at the deep beams.

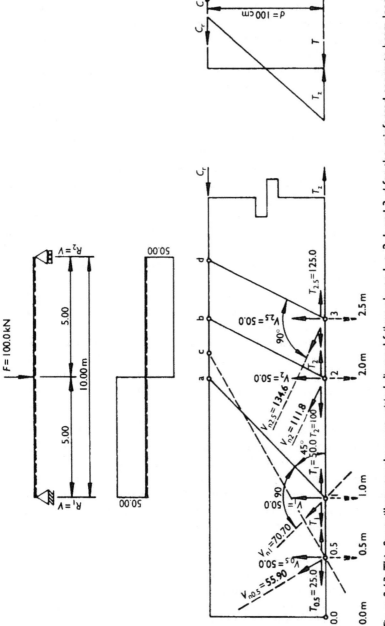

Figure 2.13 This figure illustrates the most critical diagonal failure planes 1–a, 2–b and 3–d for the reinforced concrete beam loaded by the concentrated load. The stirrups must be perpendicular to the cracking areas, which conditioned the action of the cracking forces along the stirrup (bar) itself.

3   As the tensile forces on the left side of this point become decreasingly smaller than the vertical shear forces, the cracking resultant V_n causes an increasingly greater cracking plane, with the decrease of the angle toward the horizontal, as illustrated by the line 0.5–c in Figure 2.12. In other words, the angle of the cracking plane to the left of Point 1 is decreasing, while it is increasing to the right of this point.

4   This is a free failure plane, since there is no fixity on the left side conditioned by potential shear forces A and B because the cracking plane almost disappears behind Point 1. Also, these forces (A and B) are almost too small to have an influence on the main failure plane 1–a. Thus, practically speaking, forces A and B in the cracking plane 0.5–c have no influence on the cracking plane 1–a.

5   The other possible crack could be at Point 2, where potential cracking is blocked by the fixity of configuration 2–b–a–1 in Figure 2.12, since this portion has the tendency to slide upward along the force A_2, while force B_1 (plane 1–a) is acting, along a slope downward. This counteracts the sliding tendency of the left portion of plane 2–b to slide upward. Also, a portion of the resultant V_n, responsible for causing the cracking in Point 2, is much smaller than the cracking resultant V_n in Point 1 for the uniform load, when the forces absorbed by the horizontal reinforcement are excluded (see Section 2.6.3.1 I and II of this chapter).

6   The shear forces A and B, acting in the planes left and right from the crack due to the uniform load, are strongest in the diagonal located at an angle below 45°. This is the plane 1–a.

7   In addition, the vertical shear forces V are the most influential, with the biggest impact on safety, just in the area between Points 1 and a. Thus, each of these forces in the shear plane 1–a is resolved into two forces; one force that is perpendicular, V_p, and the other force that is parallel to the plane 1–a (A'), as illustrated in Figure 2.14.

It should be noted here that, for each possible additional crack, the failure angle increases until the angle becomes 90° in the critical cross section. This conforms to everyday observations of cracking in concrete beams.

As can be seen from Figure 2.12, the possible failure lines 2–b and 3–d, at a distance approximately equal to one-half of the beam depth, are nearly parallel. Thus, stirrups can be calculated only for the portion of the beam length equal to the depth of the beam. At concentrated loads, this calculation is important only for the determination of the stirrup's slope, which will follow the direction of the resultant V_n. It is logical that the stirrups will carry only the stresses that the concrete is not able to absorb, as has been the practice until now.

The presence of a concentrated force (load) has an important influence on the occurrence of cracks where, by its presence, it can force a crack already initiated at some point, to separate from its natural cracking angle, and become oriented instead in the direction of the given force. This is the result of the combination of cracking stresses, such as support force R, and the cracking stresses of the given force, or the uniform load and the concentrated load (Figure 2.8).

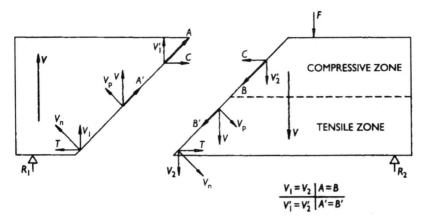

$$\frac{V_1 = V_2 \quad | A = B}{V'_1 = V'_2 \quad | A' = B'}$$

Figure 2.14 Graphical presentation of internal active forces in a flexural bending element. Besides the sliding forces caused by a combination of vertical shear forces V with compressive forces C, sliding forces A' and B' are developed in a tensile zone. This is a result of rectangular resolution of vertical shear forces V into one force perpendicular to the cracking plane (V_p) and another parallel to the cracking plane (A'). As a result, active sliding cleavage forces are developed in the compressive and tensile zone where one pair (AA') acts upwardly while another pair (BB') acts downwardly, simulating wedge action in opposite directions.

On shallow beams, the concentrated load creates its own cracking line at an angle of 45°. This is a consequence of the law of physics, namely, that critical cracking stresses, due to concentrated loads, spread out at angles below 45°. In experiments for pull-out portions of bars having been cast into the concrete mass, it can be seen that a cone is formed with an external mantle under the angle of 45°. This is generally why stirrups should be placed under each concentrated load, preventing the formation of such cracks, in conformity with the real magnitude of the stresses due to this force. Of course, until now, no importance has been attached to this because the phenomenon of diagonal failure has never been understood (see also part B of Section 2.4.4.1).

Since tensile forces do not exist above the neutral axis, it is not necessary to consider stirrups for diagonal failure above the neutral plane; to be anchored, in this case, the legs must be bent to lie in the neutral plane. But, from a practical point of view, to control sliding shear forces A and B, stirrups should extend to the top of the beam in order to simultaneously control sliding shear forces $(A \text{ and } B)$ and to serve as support reinforcement for the flexural bars in the compression zone.

At the end of this discussion, it should be emphasized very clearly that the concept, namely that all cracks are parallel in a loaded concrete beam – "The inclination of the diagonal crack is generalized by assuming it forms at an angle θ with the axis of the member,"[1] ... "This angle is constant throughout the beam, except at support and midspan where the local effect, known as the 'fanning' of the concrete struts, occurs"[34] – is fundamentally wrong! No testing example has ever been developed showing that such cracks are indeed parallel, as has been assumed. On the contrary, this angle of cracking is different at different locations on the beam as a result of the change in the magnitude of the forces acting at any location on the loaded member, as illustrated by Figures 2.12 and 2.13. But any possible cracking is almost completely predictable at any location on the beam for uniform or concentrated loads, as is explained in Section 2.6. To conclude, an incorrect assumption could never lead to the safe design of a concrete structure against shear forces in a reinforced concrete (RC) member.

2.6.1.1   Mathematical analysis of the internal active forces

The determination of the internal active compression, tensile and vertical shear forces which control the safety of the beam is illustrated in Figures 2.12 and 2.13. Internal resisting forces are equal, in magnitude, to internal active forces but they are oppositely oriented, as shown in the same figure.

The symbols used in the following calculation are:

M which denotes the internal active bending moment for the given cross section;
T the internal active tensile force for the given cross section;
C the internal active compression force for the given cross section;
d the distance (arm) between internal active compression force C and internal active tensile force T;
V the internal active vertical shear force for the given cross section; and
V_n the internal active resultant punch shear force for a given cross section.

(a) Beam subjected to uniform load
Internal active bending moment at this location:

$$M_{0.5} = 50.00 \times 0.5 - 10.00 \times 0.25 \times 0.5 = 23.75 \, \text{kN m} \quad (2,375 \, \text{kg m}).$$

Internal tensile force at this location:

$$T \times d = 23.75; \qquad T = 23.75/1 = T_{0.5} = 23.75 \, \text{kN} \quad (2,375 \, \text{kg}).$$

Internal active shear force at this location:

$$V_{0.5} = 50.00 - 10.00 \times 0.5 = 45.00 \, \text{kN} \quad (4,500 \, \text{kg}).$$

Internal active resultant punch shear force at this location:

$$V_{n0.5} = \sqrt{(45.00^2 + 23.75^2)} = 50.88\,\text{kN} \quad (5,088\,\text{kg}),$$

$$M_{0.9} = 50.00 \times 0.9 - 10.00 \times 0.45 \times 0.9 = 40.95\,\text{kN m} \quad (4,095\,\text{kg m}),$$

$$T_{0.9} = 40.95/1 = 40.95\,\text{kN} \quad (4,095\,\text{kg}),$$

$$V_{0.9} = 50.00 - 10.00 \times 0.9 = 41.00\,\text{kN} \quad (4,100\,\text{kg}),$$

$$V_{n0.9} = \sqrt{(41^2 + 41^2)} = 58.00\,\text{kN} \quad (5,800\,\text{kg}).$$

$$M_{1.9} = 50.00 \times 1.9 - 10.00 \times 1.9 \times 0.95 = 77.00\,\text{kN m} \quad (7,700\,\text{kg m}),$$

$$T_{1.9} = 77.00/1 = 77.00\,\text{kN} \quad (7,700\,\text{kg}),$$

$$V_{1.9} = 50.00 - 10.00 \times 1.9 = 31.00\,\text{kN} \quad (3,100\,\text{kg}),$$

$$V_{n1.9} = \sqrt{(77.00^2 + 31.00^2)} = 82.80\,\text{kN} \quad (8,280\,\text{kg}).$$

$$M_{2.2} = 50.00 \times 2.2 - 10.00 \times 2.2 \times 1.1 = 85.80\,\text{kN m} \quad (8,580\,\text{kg m}),$$

$$T_{2.2} = 85.80/1 = 85.80\,\text{kN} \quad (8,580\,\text{kg}),$$

$$V_{2.2} = 50.00 - 10.00 \times 2.2 = 28.00\,\text{kN} \quad (2,800\,\text{kg}),$$

$$V_{n2.2} = \sqrt{(85.80^2 + 28.00^2)} = 88.58\,\text{kN} \quad (8,858\,\text{kg}).$$

(b) Beam subjected to concentrated load

$$M_{0.5} = 50.00 \times 0.5 = 25.00\,\text{kN m} \quad (2,500\,\text{kg m}),$$

$$Cd = Td = 25.00 \times 1; \quad T = 25.00/1 = T_{0.5} = 25.00\,\text{kN} \quad (2,500\,\text{kg}),$$

$$V_{0.5} = 50.00 = 50.00\,\text{kN} \quad (5,000\,\text{kg}),$$

$$V_{n0.5} = \sqrt{(25.00^2 + 50.00^2)} = 55.90\,\text{kN} \quad (5,590\,\text{kg}).$$

$$M_1 = 50.00 \times 1 = 50.00\,\text{kN m} \quad (5,000\,\text{kg m}),$$

$$T_1 = 50.00/1 = 50.00\,\text{kN} \quad (5,000\,\text{kg}),$$

$$V_1 = 50.00 = 50.00\,\text{kN} \quad (5,000\,\text{kg}),$$

$$V_{n1} = \sqrt{(50.00^2 + 50.00^2)} = 70.70\,\text{kN} \quad (7,070\,\text{kg}).$$

$$M_2 = 50.00 \times 2 = 100.00\,\text{kN m} \quad (10,000\,\text{kg m}),$$

$$T_2 = 100.00/1 = 100.00\,\text{kN} \quad (10,000\,\text{kg}),$$

$$V_2 = 50.00 = 50.00\,\text{kN} \quad (5,000\,\text{kg}),$$

$$V_{n2} = \sqrt{(100.00^2 + 50.00^2)} = 111.80\,\text{kN} \quad (11,180\,\text{kg}).$$

$$M_{2.5} = 50.00 \times 2.5 = 125.00\,\text{kN m} \quad (12{,}500\,\text{kg m}),$$

$$T_{2.5} = 125.00/1 = 125.00\,\text{kN} \quad (12{,}500\,\text{kg}),$$

$$V_{2.5} = 50.00 = 50.00\,\text{kN} \quad (5{,}000\,\text{kg}),$$

$$V_{n2.5} = \sqrt{(125.00^2 + 50.00^2)} = 134.60\,\text{kN} \quad (13{,}460\,\text{kg}).$$

2.6.1.2 Design of details and the nature of reinforcement by stirrups

Field from Point 0 to Point 1

As presented in Figures 2.12 and 2.13, in the field from Point 0 to Point 1, stirrups are necessary in slender beams only against pure shear (vertical shear), while in deep beams only support stirrups are necessary. It can be said that it is sufficient to have one-third of the tensile bars bent in order to control the possible negative moment of the supporting portion of a simply supported beam thus increasing the resistance against pure shear between the support and the critical shear cross section. In the slender (shallow) continuous beam, cracks appear in the form of a fan, where the main (punching) angle is 45° and increases towards the vertical. This fan phenomenon can be seen clearly at the continuous beam's failure, indicating the tendency of the concentrated load to penetrate by punching out through the mass of the slender beam. Here, it is obvious that the term "fanning" is the result of the attempt to somehow explain this unknown phenomenon, applying recognized terms for the explanation of the diagonal failure. Lack of knowledge is replaced by fancy terms such as "complicated phenomenon" or "matters are being complicated," which are euphemisms for "we do not understand this" or, in simplest terms, "the problem exists, because we do not understand the problem." So we come to the following conclusion – in order to prevent such cracks, it is necessary to allow the cracking forces to lie in the stirrups themselves. In other words, stirrups must be perpendicular to the given cracks.

The main reasons why diagonal cracks rarely appear in this field (0 to 1) are the following:

a   The cracking resultant to the left of the critical shear cross section (where vertical shear forces are equal in size to the horizontal tensile forces) is less than the critical resultant at a distance from the support approximately the depth of the beam. This is the case, even though the vertical shear force for the uniform load increases toward the support. That is why a diagonal crack cannot appear, since diagonal cracking is not conditioned by vertical shear force, but instead, by the cracking resultant V_n.

b   As previously stated in our study, horizontal reinforcement absorbs the main portion of the tensile force T. Thus, for diagonal tension only the portion equal to the vertical shear force remains for the given cross section. And, since the horizontal tensile forces T (behind the vertical diagonal critical section) are smaller than the vertical shear forces V, diagonal tension and horizontal tensile forces T, in fact, do not exist (since all of them are absorbed by the horizontal reinforcement). This means that to the left of the diagonal critical section, the action of the vertical shear force V, is resisted by the entire area of the concrete section with compression, tensile and every other reinforcement penetrating into this section. That is why the danger of the so-called diagonal tension does not exist in that area.

Here we need to envision punch shear in relation to the support itself, where a lack of such reinforcement can cause such a failure in a shallow member. Punch shear, although at angle below 45 from the support itself, has nothing in common with the so-called diagonal failure due to the combination of horizontal tensile forces T with vertical shear forces V to the right of the critical diagonal tension. So, here reinforcement is necessary only if the beam is shallow, because the concrete itself cannot prevent the punch shear of the support. In this case some reinforcement must exist to prevent this punching.

Field from Point 1 to Point 2
It is also clear from Figures 2.12 and 2.13 that the field from Point 1 to Point 2 is the most critical for the given beam, since the shear resultants here are oriented under the natural cracking angle of 45 , which is characteristic of each free punch shear (cracking). Horizontal tensile reinforcement contributes, in a very small measure, to the prevention of the appearance of cracks, as a result of the tendency of the left portion to move upward, and of the right one to move downward. But, if this reinforcement is placed in horizontal layers up to the neutral axis itself, then the fixing (anchoring) of bars left and right of the crack would prevent the motion of the left portion upward and the right one downward. Thus, a crack may not appear and, if it does appear, it cannot develop further. But, such a solution creates a very irrational use of reinforcement since the rule is that the most rational protection of the concrete element against diagonal cracks be used, if the forces causing these cracks act along the reinforcing bar (stirrup) itself. That means that this field requires stirrups to be placed in the slope of 45°. It follows from this that, for Figure 2.12, stirrups will be calculated for the force value V_n of 58.00 kN (after the subtraction of the force absorbed by the concrete), while for Figure 2.13, stirrups would be calculated for the force value V_n of 70.70 kN (after subtraction of the force absorbed by the concrete).

Field from Point 2 to Point 3
Here the motion (displacement) of the left portion from Point 2 upward and of the right portion downward becomes difficult, as presented in the seven points in Section 2.6.1. This means that the possible displacement upward is a function of the given vertical shear force for the given point, 31 kN for the uniform load (Figure 2.12), and 50 kN for the concentrated load. In other words, the cracking resultant for the uniform load $V_{n1.9} = \sqrt{(77.00^2 + 31^2)} = 82.80$ kN, and for the concentrated load $V_{n2} = \sqrt{(50^2 + 100.00^2)} = 111.80$ kN. After the subtraction

of the force that concrete can absorb, stirrups will be charged with the rest of the plane 2–b, where forces will be acting along the stirrups. As already related, if horizontal reinforcement does not exist, the calculation of stirrups for preventing diagonal failure would be 82.8 kN for the uniform load, and 111.80 kN for the concentrated load, as illustrated in Points 2 in Figures 2.12 and 2.13. This method will be used for each following field.

However, it is necessary to point out that if concentrated loads exist in the proximity of the 45° angle at the top of the compressed zone or at the bottom of the tensile zone, then the slope of the same stirrups will be combined. This is in order to cover the shear stresses due, concurrently, to regular loading and to an additional concentrated load.

It is necessary to remark that sometimes the term "principal shear stresses" is used for diagonal stresses. This term is not well-chosen, since such stresses do not exist. Main (principal or normal) stresses of compression or tension exist on one side and shear stresses or diagonal tension on the other side. In particular, the question is: If these are the principal shear stresses, then which are the secondary shear stresses? In fact, the shear stresses can be calculated from principal stresses, by using trigonometric formulas. This is also the case with torsion. But nowhere is it noted that principal shear stresses can be obtained from principal compressive and tensile stresses, since secondary shear stresses do not exist.

2.6.2 Inverted T beam for bridge structures and the nature of its stirrups

In order to emphasize the importance of this discussion, we will quote from published material.

> The loads that are introduced into the bottom rather than into the sides or the top of the web of an inverted T beam impose special problems, *which are not dealt with in existing structural codes*. Hence, a general analytic solution for strength of an inverted T beam that is versatile enough for all possible load cases and simple enough for design office application does not seem to be within reach at the present time. We have, therefore, resorted to empiricism supported by a rational interpretation of test results in order to develop design criteria for inverted T beams.[35] (Here, we have used italic for emphasis.)

For every problem there are, in general, multiple solutions that follow from a complete understanding of the problem. Consequently, the suggestion of taking the load from the stringers directly by the flexural tensile zone of an inverted T beam (as suggested by the authors) is probably the last possible solution which should be applied, for the following reasons:

(1) Generally, the depth of the flange is designed for punch(ing) shear as the authors stated – "the flange must be deep enough to avoid punching shear

weakness"[35]—and not for a flexural compression and tension. As a result, the flange is much higher than it needs to be.

(2) As has been stated in paragraph A of Section 2.4.4.1, punch shear strength is not a reliable parameter because, for the same quality of concrete, punch shear tests performed by a pullout test for an unloaded member give one value and the same pullout test, taken from a flexurally bent member, gives a completely different value. It is much higher if the pullout test is performed from the neutral axis of the bent member upwardly (opposite to bending forces), and much lower if the same pullout test were performed downwardly, from the tensile zone only, parallel to bending forces. We could state, as a rule, that the larger the flexural tension of the member, the smaller the resistance to punch shear.[14–16]

For these reasons, it is impossible to ever depend on the strength of concrete against punch shear in a flexural member (with the load located in the tensile zone). This is true specifically in a situation where it is possible for a sudden dynamic impact to occur, such as on a bridge, where exposure to sudden loads is quite common.

To be reliable, the punch shear test must be performed for the worst flexural bending with a pullout test *from the level of the upper portion of the bracket* (flange) downwardly. In this case, the yield of the punch shear would drop down to approximately one-tenth of the yield obtained from the unloaded member. The reason for this is that we need punch shear resistance under the worst flexural tension in the tensile zone itself and not the yield strength for unloaded members (as the authors' value is attained and suggested to us[35]) because the load is located in the tensile zone.

(3) A bracket, as a structural member, is very unsafe considering the location and disposition of reinforcement for the following reasons:

a   A bracket (flange) as a cantilever is not designed as a cantilever because it lacks a real reinforcement to serve as such. This is the case, for example, with a corbel where all forces are transferred due to reinforcement in the main body of the beam or column.

b   A cantilever in the tensile zone does not have any compressive concrete zone to increase its resistance to punch shear cracking. As a result, safety is based on very limited resistance to punch shear (because of its location in the tensile zone), and possible transferring of vertical load by vertical stirrups via bond forces between the cement gel and the stirrups. Unfortunately, the rule here is also that larger flexural tension stipulates smaller bond resistance.

Even if stirrups started to carry some load from the flanges, by increasing the flexural tensile stresses, bonding stresses decrease so the only real safety is a function of the punch shear resisting stresses. Yet, if somehow the same stirrups could be loaded by the stringers directly (by hanging), almost all of the problems would be resolved and the member (T beam) would fail due strictly to flexural bending. Even if the entire resisting bond stresses collapsed in the tensile zone, the stirrups will transfer their loads to horizontal reinforcement in a compressed zone (or at the top of the member) and failure would be prevented.

(4) Torsion is enormous and, yet, is located exclusively in the tensile zone. The rule here would also be that the larger the flexural bending stresses the smaller the resistance to torsional stresses.

A much better solution to the above dilemma could be achieved if stringers were literally hung up to the top of the compressive and tensile zone by transferring the load from the flange directly to the top of the beam,[36] as illustrated in Figures 2.15 and 2.16. By doing this, the problem of punch(ing) shear would be eliminated almost totally because the compressed zone is much more resistive to punch shear. Horizontal reinforcement of the flanges (brackets) (without any supports as the authors suggested) will be supported by the top of the beam via "hangers." The entire design would be as that for concentrated loads hung directly at the top of the compressive zone through corresponding horizontal bars without any further concern about punch shear in the flange.

A similar solution could be developed by transferring the stringers' load directly at the top of the flexural tensile zone (in the field of negative moment) where the entire cross section of the beam will react against any type of shear failure. Resistance to punch shear, in this case, will be strengthened by the compressive zone at the bottom of the beam. The higher the compressive stresses, the higher the resistance to punch shear for any individual stringer to punch out a portion of the beam and cause failure.

For a very dense horizontal reinforcement, as shown in authors' Figure 13(c),[35] the horizontal reinforcement will be added under the existing dense layers in order to take over the entire load from the stringers via "hangers" (special stirrups designed to do such a job).

It should be clearly emphasized that by eliminating punch shear, frictional shear, vertical shear and horizontal shear in the bracket will be eliminated, and by modifying torsional problems, as suggested and explained below, the real failure of the inverted T beam would be ductile flexural failure.

With such a design and treatment of any inverted T beam, there would be little or no justification for the authors' statement that "the longitudinal and lateral bending of the flange of an inverted T beam produces a very complex stress distribution in the flange."[35] Also, the main problem, as the authors state, "the tests on an inverted T beam have shown that hanger failure cracks occur at the junction of the web and the flange,"[35] will be eliminated completely.

Probably the simplest transferring (hanging) reinforcement from a stringer to the top of the main beam would be a simple stirrup[36] (with the corresponding cross section, calculated to take over the entire stringer's load at the top of the beam. Such a hook-up of stirrups will securely prevent any possible pullout and bond failure (see Figure 2.15).

Irrespective of the type of reinforcement applied (the reinforcement of the cited authors' or ours), one completely new feature should be introduced: metal pads

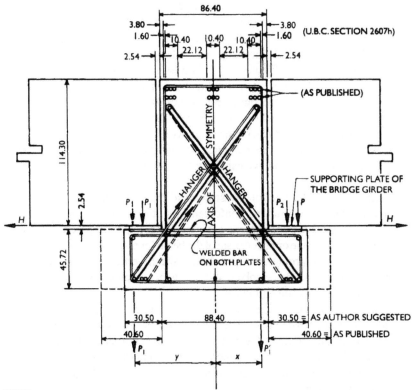

NOTE:

H, THE HORIZONTAL FORCE OF THE BRIDGE GIRDER (OR DISPLACEMENT DUE TO SEISMIC FORCES);
P, CONCENTRATED LOADS LOCATION, AS PUBLISHED; P_1 AND P_2 = CONCENTRATED LOADS
LOCATION, ACCORDING TO OUR SUGGESTION; P_1 = THE BRIDGE BEAM LOAD TRANSFERRED TO
THE TOP OF THE BEAM; y = ECCENTRICITY APPROXIMATELY 660 mm AS PUBLISHED; x = ECCENTRICITY
APPROXIMATELY 356 mm ACCORDING TO OUR SUGGESTION, REDUCED BY 46 PERCENT.

Figure 2.15 The figure shows that the bridge girder loads are transferred to the top of
the beam, using "hangers" (special stirrups). By such manipulation with forces
in the beam, the punch shear will be eliminated completely, as well as the
frictional shear, and the beam can be treated as any beam with the load on the
top of the beam. Also, the eccentric load from the bridge girders (force P_1)
will be transferred to the top of the beam, as force P_1, replacing the action of
the force P_1 with reduced eccentricity and reduction torsion.

should be placed under any stringer with or without rubber pads and by a corresponding bar be welded to the opposite pad (through the web of the main beam)
so that horizontal forces (developed by shrinkage of the stringer, earthquake, or
any other possible movement) would be arrested, so that the eventual movement
of the stringer and their horizontal forces could never lead to possible punch shear

cracking anywhere under the pad or between the surface of the flange and the vertical surface of the web (Figures 2.15 and 2.16).

If our concept is used, the problem of torsion would be modified considerably because the hanging force, say left force P_1, will cause another force P_1' at the opposite side of the vertical symmetry line and, by so doing, eccentricity would be greatly decreased (see Figure 2.15). In other words, when the left stringer induces its eccentric force P_1, at the left bracket, simultaneously, this force would be transferred to the top of the right-hand side of the beam, and new force P_1' will counteract force P_1 at a smaller distance from the vertical symmetry line with obviously smaller torsion. Also, the eccentricity of the loading could be modified further by applying a little more complicated "hanger" stirrup (see Figure 2.16).

From the above discussion, we can draw the following primary conclusions:

a   By transferring the load of the stringer to the top of the beam, using the same peripheral longitudinal reinforcement of the flange (bracket) as bearing reinforcement, and by transferring the horizontal forces, imposed on the plate by the stringer, to the opposite plate of the beam (or opposite side of the column for a corbel), all problems of punch(ing) shear, vertical shear of the flange, frictional shear, would be solved and the "very complex stress distribution in the flange"[35] would be eliminated.

b   If, somehow, residual punch shear stresses occur for any possible reason, or unforeseen, sudden dynamic impact at the stringer occurs, diagonally installed hanging stirrups would be the best possible protection against the punch shear developed because they cross almost perpendicularly at the location of the possible occurrence of punching shear (at the corner of the flange and the web of the main beam).

c   Using our solution for the loading of an inverted T beam, the problem of torsion will be greatly modified by substantially decreasing the torsional eccentricity (approximately 46%), so the influence of the torsional effect would be much smaller. Consequently, bridge design and safety would be much improved, which is the real motivation for this discussion.

d   By applying a "hanger," as shown in Figure 2.16, the load from the stringers could be transferred directly to the symmetry line of the beam.

2.6.2.1 Failure due to radial tension and its stirrups

This kind of failure of beams without reinforcement comes to full expression in wooden laminated curved beams (arch like) loaded in flexure. In the literature, such a failure is termed "failure due to radial tension." In an ordinary wooden beam, in the proximity of the critical flexural section, the failure would occur due to compression and tension; yet, at the raised tensile zone, such a failure cannot occur unless the bending line does not cross under the plane connecting these two supports, while its forces pushing outward are eliminated. Since failure due to compression and tension cannot occur, it is quite natural that the punch shear

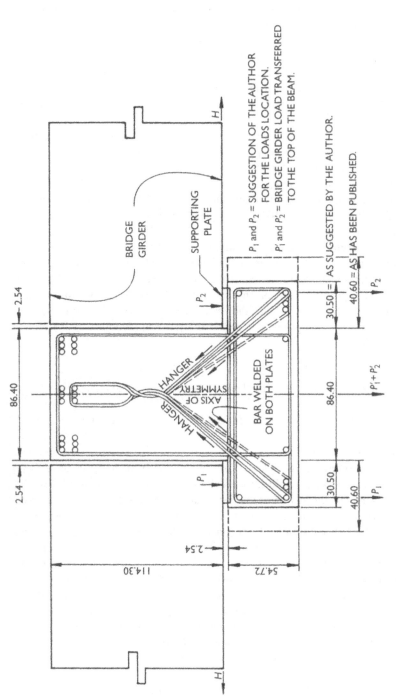

Figure 2.16 By a more complicated "hanger" (stirrup), the load of the bridge girder can be transferred directly to the top of the beam in the axis of symmetry.

failure develops in the proximity of the critical section, where the resultants of the cracking forces are maximal. It is clear that this is the function of the permanent increase of tensile force toward the critical section, while the vertical shear forces are smaller and smaller (or constant, depending on the loading). To emphasize, it is obvious that tensile forces in the RC beam are controlled by tensile reinforcement, while in this case, such a control does not exist. Thus, the diagonal failure comes to full expression where the punch shear force resultants are maximal, as illustrated in Figures 2.12 and 2.13.

The prevention of such failure is realized by metal bars being built into the beam (or outside of it). These bars are placed perpendicularly to the action of the punch forces, so that the cracking forces act along the bars themselves.

Similarly, as in the concrete, in order to provide security against shear forces by a greater stress state, the American Institute for Timber Construction, in its manual (pages 5–230, third edition, 1985), recommends the use of "a radial reinforcement sufficient to resist the full magnitude of radial tension force." This corresponds to the built-in stirrups in the concrete beam.

It is interesting that the existing calculation control is based on experimental logic, without a real understanding of the stresses and the forces creating them. Our method applies the laws of physics, where the forces causing such cracks and failures are seen clearly. Thus, they can be easily controlled by corresponding reinforcement. The real problem is eliminated when the problem itself becomes understandable.

2.6.3   Shallow beams: an analytical approach to our theory

2.6.3.1   Reinforcement calculation for the control of diagonal failure

Existing (classical) method
Consider the following example: a slab with the web executed as an ordinary single-span beam, with a uniform load along the beam, $q = 40.00\,\text{kN/m}$. The span is 8.00 m, the dimensions of the slab with the web are already calculated: the statical height $h = 50\,\text{cm}$, the arm of internal forces $z = 46\,\text{cm}$, the web width, 25 cm. The beam is constructed from concrete $300\,\text{kg/cm}^2$, the reinforcement for shear stresses steel GA 240/360 ($\sigma a = 140\,\text{MPa}$). The slab thickness is 10 cm. It is necessary to calculate the reinforcement for shear stresses (diagonal tension). The reinforcement for flexural bending is the following: $4 \oslash 25$ and $5 \oslash 20$; a portion of shear reinforcement will be achieved by bending one half of the area of the existing bars below 45°; in this case $5 \oslash 20$ to absorb shear stresses and for absorbing the ever-present negative moment of undefined magnitude (under a bending condition, the supported portion of the beam becomes a cantilever).

Vertical shear force at the support:

$$V_{max} = ql/2 = 40.00 \times 8.00/2 = 160.00\,\text{kN} \quad (16{,}000\,\text{kg}).$$

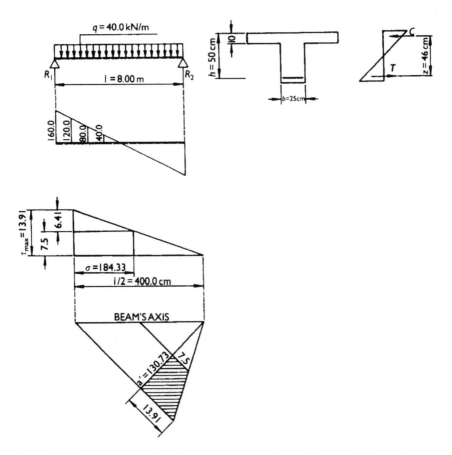

Figure 2.17 Determination of reinforcement to control diagonal "tension" failure using the classical method.

Maximal stresses of diagonal tensile forces:

$$\tau_{max} = V_{max}/bz = 0.160 \, \text{MN}/0.25 \times 0.46 = 1.391 \, \text{MPa} \quad (13.91 \, \text{kg/cm}^2).$$

This is the case when all the diagonal tension is absorbed by the reinforcement. The length of the beam that must be secured against diagonal tension is:

$$a : 0.641 = 4.00 : 1.391; \quad a = 0.64 \times 4.00/1.391 = 1.84 \, \text{m}.$$

For the calculation of the tensile force the length a' is:

$$a' = a/\sqrt{2} = 1.84/1.4 = 1.31 \, \text{m}.$$

The total shear force is:

$$Ss = (0.75 + 1.391)a'b/2 = 1.07 \times 1.31 \times 0.25 = 0.3497 \, MN$$

$$(34,970 \, kg).$$

Required reinforcement:

$$Az = Ss/\sigma a = 0.3497/140 = 0.00249 \, m^2 = 25.00 \, cm^2.$$

Bent half of the main reinforcement $5 \oslash 20 = 15.71 \, cm^2$.

$$\text{Stirrups: } 24.99 - 15.71 = 9.28 \, cm^2 \approx 25.00 \, cm^2.$$

Use 10 stirrups with $10.06 \, cm^2$.

a   U-shaped stirrups $10 \oslash 8 = 10.06 \, cm^2$, at 18.43 cm; or
b   U-shaped stirrups $13 \oslash 7 = 10.00 \, cm^2$, at 14.18 cm.

New method
This method is based on the efficiency conditioned by the combination of the internal active forces of compression and tension with internal vertical active shear forces. There are two cases to be distinguished: the deep-beam case and the shallow-beam case.

In deep beams, the cracking planes lie perpendicular to the resultant of the real punching forces, where cracking slopes have a tendency to create an angle greater than 45° just behind the first field. This cracking direction is well followed by the so-determined potential cracking planes, as was presented in the section on deep beams.

In shallow beams, where reinforcement against diagonal cracks is required, the cracking angle is almost 45°, lying in the first one-third of the beam, left or right from the support. The second case is elaborated here.

I   Reinforcement necessary for the critical section (plane 0.43–a)
The critical section is located where internal horizontal tensile forces T are equal in value to the internal vertical shear forces V. In this case it is calculated that these forces are equal in value at a distance of 0.43 m from the left support (or to the right side from the support).

According to the shear diagram, determination of the location of the critical shear cross section is based on the fact that $C = T = V$, namely:

$$(16,000 \cdot x - 4,000 \cdot x \cdot x/2)/0.46 = 16,000 - 4,000 \cdot x,$$

$$x^2 - 8.92 \cdot x + 3.68 = 0; \qquad x_1 = 0.434 \, m;$$

$$x_1 \text{ (rounded off)} = 0.43 \, m.$$

Magnitude of internal forces at critical shear cross section:

$M_{0.43} = 160.00 \times 0.43 - 40.00 \times 0.43 \times 0.215 = 65.10\,\text{kN m (6,510 kg m)}$,

$C_{0.43} = 65.10/0.46 = T \approx 141.52 \approx 142.00\,\text{kN (14,200 kg)}$,

$V_{0.43} = 160.00 - 40.00 \times 0.43 \approx 142.80 \approx 142.00\,\text{kN (14,200 kg)}$,

$V_{n0.43} = 142.00 \times 1.41 = 200.22\,\text{kN (20,022 kg)}$.

Reinforcement necessary for the plane 0.43–a

$Ar = V/\sigma_a = 0.20022/140 = 0.00143\,\text{m}^2 = 14.3\,\text{cm}^2$.

Diagonal tension stresses for the plane 0.43–a

$\tau_{0.43} - a = 0.20022/0.705 \times 0.25 = 1.136\,\text{MPa (11.36 kg/cm}^2)$.

In this field the reinforcement absorbs all the stresses, the concrete not being charged with any contribution, since this is the real critical area of diagonal failure.

Use 10 stirrups with $14.14\,\text{cm}^2$.
U-shaped stirrups $10 \oslash 10 = 14.14\,\text{cm}^2$ at 5.2 cm.

II Reinforcement necessary for the plane 0.95–b
In this zone the concrete will be charged with those stresses that do not require reinforcement according to the European standard. As was applied here, the $300\,\text{kg/cm}^2$ concrete with $7.5\,\text{kg/cm}^2$ for shear, is allowed to charge the concrete with $4.5\,\text{kg/cm}^2$.

$M_{0.95} = 160.00 \times 0.95 - 40.00 \times 0.95 \times 0.475 = 133.95\,\text{kN m (13,395 kg m)}$,

$C_{0.95} = T_{0.95} = 133.95/0.46 = 291.19\,\text{kN (29,119 kg)}$,

$V_{0.95} = 160.00 - 0.95 \times 40.00 = 122.00\,\text{kN (12,200 kg)}$.

As the horizontal reinforcement absorbs the main portion of the tensile force T, the portion equal to the vertical shear force at the section 0.95–a remains for the diagonal tension: that is, only the portion of 122 kN. The necessary punch shear reinforcement for this section is:

$V_{n0.95} = 122.00 \times 1.41 = 172.02\,\text{kN (17,202 kg)}$.

Here the concrete absorbs the force

$Sb = 0.705 \times 0.25 \times 0.45 = 0.07931\,\text{kN (7,931 kg)}$.

The reinforcement absorbs the force

$Sa = 172.02 - 79.31 = 92.71\,\text{kN (9,271 kg)}$.

Necessary shear reinforcement for a given plane 0.95–b:

$$Az = 92.71/14 = 6.62 \, cm^2.$$

Use 9 stirrups with $6.92 \, cm^2$.
U-shaped stirrups $9 \oslash 7 = 6.92 \, cm^2$ at 5.55 cm.

III Reinforcement necessary for the plane 1.45–c

$M_{1.45} = 160.00 \times 1.45 - 40.00 \times 1.45 \times 0.725 = 189.95 \, kN \, m \, (18,995 \, kg \, m).$

$C_{1.45} = T_{1.45} = 189.95/0.46 = 412.93 \, kN \, (41,293 \, kg).$

$V_{1.45} = 160 - 40.00 \times 1.45 = 102.00 \, kN \, (10,200 \, kg).$

This is the same as in field 0.95–b: for diagonal tension, there remains the horizontal force T, equal to the vertical force, $T = V$, so that the punch shear force is

$$V_{n1.45} = 102.00 \times 1.41 = 143.82 \, kN \, (14,382 \, kg).$$

The concrete absorbs the force

$$Sb = 0.705 \times 0.25 \times 0.45 = 0.07931 \, MN \, (7931 \, kg).$$

The shear reinforcement absorbs the force

$$Sa = 143.82 - 79.31 = 64.51 \, kN \, (6451 \, kg).$$

Necessary shear reinforcement $Az = 64.51/14 = 4.61 \, cm^2.$

Use 9 stirrups with $5.10 \, cm^2$.
U-shaped stirrups $9 \oslash 6 = 5.10 \, cm^2$ at 5.55 cm

IV Reinforcement necessary for the plane 1.95–d

$M_{1.95} = 160.00 \times 1.95 - 40.00 \times 1.95 \times 0.975 = 235.95 \, kN \, m \, (23,595 \, kg \, m),$

$C_{1.95} = T_{1.95} = 235.95/0.46 = 512.93 \, kN \, (51,293 \, kg).$

Vertical shear force:

$$V_{1.95} = 160.00 - 40.00 \times 1.95 = 82.00 \, kN \, (8,200 \, kg).$$

Punching shear force:

$$V_{n1.95} = 82.00 \times 1.41 = 115.82 \, kN \, (11,582 \, kg).$$

The concrete absorbs the force

$$Sb = 0.705 \times 0.25 \times 0.45 = 0.07931 \text{ MN (7,931 kg)}.$$

The shear reinforcement absorbs the force

$$Sa = 115.62 - 79.31 = 36.31 \text{ kN (3,631 kg)}.$$

Shear reinforcement Az, necessary to absorb this force is

$$Az = 36.31/14 = 2.59 \text{ cm}^2.$$

Use 7 stirrups with 2.74 cm².
U-shaped stirrups $7 \oslash 5 = 2.74$ cm² at 7.14 cm.

2.6.3.2 Conclusion on reinforcement calculation by the new and old methods

The necessary reinforcement, calculated according to the new method is

$$14.30 + 6.62 + 4.61 + 2.59 = 28.12 \text{ cm}^2.$$

The necessary reinforcement according to the classical method is 25.00 cm².

Yet, according to the new method, more reinforcement is necessary, totalling 3.12 cm². This is irrelevant because we are following a law of mechanics and not an unproven assumption of Ritter and Morsch, "... a case of pure shear was assumed to exist."[22] Furthermore, in cracking over the neutral zone, tensile force V_n does not exist, but shear forces V_n' exist, acting parallely to the diagonal crack, as illustrated in Figure 2.8(c). This should be covered by stirrups.

It should be emphasized here that the aim of the new method is not savings in reinforcement, but rather a comprehensive calculation of reinforcement against diagonal failure and the understanding of the failure mechanism, itself. The quantity of reinforcement is strictly a function of necessity.

The length of the beam being covered against diagonal failure, according to the classical method, is 183 cm, while according to the new method it is 202 cm. But this comparison is really irrelevant, since the classical theory is only a hypothesis which, as such, does not exist. The new method is based on the fact that in the bent beam, the combination of the internal active forces of compression and tension with the internal active forces of vertical shear naturally conditions the diagonal cracks and diagonal failure.

Note 1   When this law on diagonal failure becomes accepted world-wide, then the modifications of its application will be the task of its contemporaries.

Note 2   For the field 0–0.43 of the beam, the reinforcement does not need to be calculated, since the angle of the cracking plane is rapidly decreasing toward the horizontal, so that the cracking plane penetrates into the field between 0.43 and 0.95 (planes 0.5–c in Figures 2.12 and 2.13). Here, diagonal cracks do not appear, so that, according to the United States Uniform Building Code (UBC), the calculation of shear stresses is not required; yet, constructive stirrups in all cases are to be put into this field. On the other side, stirrups are placed at an angle below 45°, so that from the point 0.43 to the left, stirrups from the right side penetrate into the field 0–0.43.

Note 3   As stated in Section 2.6.1 of this chapter the distance from the support for the critical shear cross section is approximately $0.9 \times 0.5 = 45$ cm for the uniform load, while we have calculated it to be 43 cm.

Note 4   In any flexurally bent member, two different critical cross sections exist:

a   a critical flexural cross section where the bending moment is the largest and failure is expected to occur due to tension and compression; and

b   a critical shear cross section where internal active compression, tensile and shear forces become equal in magnitude and failure is expected to occur under the action of resultant punch shear forces.

2.7 Prestressed concrete

By recognizing that diagonal cracking is caused by a combination of vertical shear forces (V_1 caused by support, and V_2 caused by external load), and flexural tensile forces (T) (as shown in Figure 3.2), it is not difficult to visualize that by prestressing one can control the tensile force T. So, by changing the magnitude of the tensile force by prestressing, one controls the ultimate value of punch shear force V_n, which causes diagonal cracking. That literally means that, if we eliminate flexural tensile forces, we can eliminate punch shear forces V_n; consequently, we can eliminate diagonal cracking and diagonal failure. Yet, if the tensile force is reduced by prestressing to such a degree that it becomes smaller than the vertical shear forces for a given cross section, then the angle of cracking, caused by punch shear resultant V_n, becomes smaller than 45° and will not cause a real threat of diagonal shear cracking. This is illustrated by Figures 2.12 and 2.13.

Furthermore, if on prestressing the tensile force remains larger than the vertical shear forces V_1 and V_2 for a given cross section of a member, then diagonal cracking will be controlled, as explained in Section 2.6.3 of this chapter and illustrated in Figure 2.18.

In conclusion, it could be stated that for diagonal cracking to be caused, tensile force T must coexist simultaneously with vertical shear forces V_1 and V_2. Therefore, if one of these three forces is eliminated (i.e. vertical shear force V_2 by

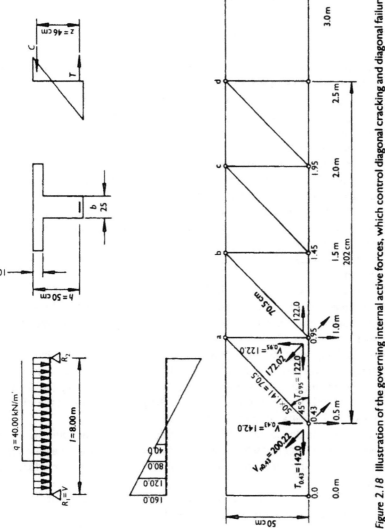

Figure 2.18 Illustration of the governing internal active forces, which control diagonal cracking and diagonal failure.

applying pure bending), punch shear force V_n (which causes diagonal cracking) as well as diagonal cracking are also eliminated.

2.8   Conclusion

Based on the above discussion and the quotations from textbooks, the following conclusions can be drawn:

1   Diagonal cracking in a flexurally bent member is caused by the resultant punch shear forces V_n as a result of the natural tendency of the support to move its portion of the beam upwardly, and the natural tendency of the external load to move its portion of the beam downwardly.

2   It becomes possible to predict, with a high degree of accuracy, punch shear cracking for a uniform load as well as for a concentrated load, and to calculate, fairly accurately, the necessary minimal web reinforcement to prevent such cracking and possible failure for a given load.

3   By applying internal active and internal resisting forces on a unit element, as given by the new law, it appears that eventual diagonal tension should be parallel to diagonal cracking.

4a   Consequently, to visualize the direction of shear forces in a flexural member, two shear diagrams must exist: one for the shear forces caused by the external load and oriented in the same direction as the load, and another for the shear forces caused by the supports and oriented in the same direction as the supports.

4b   By so doing, it will emerge immediately that the shear diagram for a beam on two supports can never be identical to the shear diagram for a fixed-end beam where shear forces at the inflection point must be oppositely oriented to each other.

5   By applying existing knowledge of shear stresses caused by bending (Saliger)[18] and resisting shear stresses (Timoshenko,[20] Winter–Nilson[23]) on a unit element, diagonal tension should occur *parallel* to diagonal cracking, while the existing theory suggested it to be *perpendicular* to the diagonal cracking, as illustrated in Figure 2.7(c).

6   For a unit element cut out from the twisting shaft, or under pure shear, real shear forces are applied which can cause diagonal tension and diagonal cracking, while for an element from a bent member, only resisting shear forces are applied.

7   Diagonal tensile stresses due to shear, as suggested by our pioneers, cannot be produced in ordinarily loaded beams (if torsion is not involved).

8   Sliding shear stresses of the existing theory are oppositely oriented to the direction of real sliding shear stresses caused by bending phenomena (see Figures 2.4 and 2.5(a), (b)).

9   Eventually, as a result of opposite directions of shear stresses (as suggested by the classical theory and shear stresses caused by bending itself), they will cancel each other, so only active punch shear forces V_n and V_n' will remain.

10   The existence of internal active and internal resisting forces in a flexurally bent member is supported in the existing literature (see references 20, 25, 37, 19, 23 and 38). Thus, their existence is an indisputable fact.

11   By proving the existence of internal active and internal resisting forces, our theory has been proved, because such forces are the essence of our theory.

12   The classical concept for the diagonal tension theory is based on pure assumption, as shown in the textbook by Wang–Salmon.[22]

13   As shown in Figure 2.14(b),[23] the classical theory of diagonal tension employs resisting shear stresses (instead of active shear stresses) on a unit element, in order to prove the accuracy of the diagonal tension theory and, as such, appears to be a fundamentally false theory.

14   The existence of diagonal tension is pure assumption, i.e. it has never been proved: "During the years since the early 1900s until the 1963 Code was issued, *the rational philosophy was to reason* that in regions where normal stress was low or could not be counted on, *a case of pure shear was assumed to exist*" (Wang–Salmon, *Reinforced Concrete Design*[22] and ACI–ASCE Committee 326 (see references 1 and 3)).

15   If one cuts a unit element from a bent member with active shear stresses, as shown by Timoshenko's Figure 160(b) – forces F, Seely–Smith's Figure 128(b) – forces V, Saliger's Figure 198(a) and 198(b) – stresses τ_v and τ_h and Bassin–Brodsky–Wolkoff's Figure 11.4, then evidently diagonal tension *appears to be parallel to diagonal cracking* of the beam.

16   The existing theory for equilibrium of a free body is not provable physically or mathematically and has not yet been proved.

17   By applying our theory the diagram for the equilibrium of a free body (cut from a flexural member) becomes algebraically provable, as is shown in this work. Thus, the phenomenon of diagonal tension becomes understandable, foreseeable and completely controllable.

18   To take the load from the stringer, located in the tensile zone of the inverted T beam, into the flexural compressed zone of the same inverted T beam is, probably, the best possible solution to design bridge structures.

19   The new law of engineering mechanics is based on the existence of the internal active and internal resisting forces of compression, tension, and horizontal and vertical shear in any flexurally bent member.

20   The combined action of internal active tensile forces with internal active vertical shear forces, causes diagonal cracking in the tensile zone.

21   The combined action of internal active compression forces and internal active vertical shear forces, causes sliding shear forces in the compressed zone, acting simultaneously with diagonal cracking in the tensile zone.

22   Newton's third law manifests itself throughout the universe. The new law manifests itself exclusively and only in flexural bending and nowhere else.

23   While the new law is based exclusively on the existence of internal active and internal resisting forces, which are created in the body of a bent member,

Newton's third law is based on an external active force and the internal resisting force which tends to prevent the effect of the external force.

24 Finally, Newton's third law and our new law are two totally different laws of physics.

References

1. ACI-ASCE Committee 326, "Shear and Diagonal Tension", Proceedings, *ACI*, Vol. 59, January–February–March, 1962, pp. 3, 7, 18, 21.

2. Stamenkovic, H. (1981) "Shear and Torsion Design of Prestressed and Not Prestressed Concrete Beams", discussion, *PCI Journal* **26**, 106–107.

3. Joint ASCE-ACI Committee 426 (1973) "The Shear Strength of Reinforced Concrete Members", *Journal of the Structural Division* **70**, 1117, 1121.

4. Barda, F., Hanson, J. M. and Corley, W. G. (1977) "Shear Strength of Low-Rise Walls with Boundary Elements", *Reinforced Concrete Structures in Seismic Zones*, ACI Publication SP-53, Detroit, MI, pp. 153, 154.

5. Stamenkovic, H. "Diagonally Reinforced Shearwall can Resist Much Higher Lateral Forces than Ordinary Shearwall", Proceeding of the Eighth World Conference on Earthquake Engineering, July 21–28, 1984, San Francisco, CA, Vol. 5, pp. 589–596, Figure 1a.

6. Palaskes, M. N., Attiogbe, E. K. and Darwin, D. (1981) "Shear Strength of Lightly Reinforced T-Beams", *ACI Journal* **78**, 450–451, Figure 3.

7. Fuller, G. R., "Earthquake Resistance of Reinforced Concrete Buildings State-of-Practice in United States", Proceedings of – Workshop on Design of Prefabricated Concreted Buildings for Earthquake Loads, April 27–29, 1981, Los Angeles, Applied Technology Council, Berkeley, CA, p. 134.

8. Cravens, R. P. (1951) "Structural Requirements of Building Code", *Modern Building Inspection* 2nd edn, eds R. C. Colling and Hall Colling, Building Standard Monthly Publishing Company, Los Angeles, CA, p. 314, Figure 5.

9. Park, R. and Paulay, T. (1975) *Reinforced Concrete Structures*, John Wiley and Sons, New York, pp. 271, 276, 307.

10. Kani, G. N. J. (1969) "A Rational Theory for the Function of Web Reinforcement", *ACI Journal*, Proceedings **66**, p. 86.

11. McGuire, W. (1959) "Reinforced Concrete", *Civil Engineering Handbook*, 4th edn, editor in chief Leonard Church Urquhart, McGraw-Hill Book Company, New York, pp. 7–128, 129.

12. McCormac, J. C. (1986) *Design of Reinforced Concrete*, 2nd edn, Harper and Row, New York, pp. 190, 191.

13. Stamenkovic, H. (1978) "Suggested Revision to ACI Building Code Clauses Dealing with Shear Friction and Shear in Deep Beams and Corbels", discussion, *ACI Journal* **75**, 221–224, Figures Ca, Cb, Cc, Cd – Forces A and B.

14. Stamenkovic, H. (1980) "Design of Thick Pile Caps", discussion, *ACI Journal* **77**, 478–481, Figure b.

15. Stamenkovic, H. (1981) "Comparison of Pullout Strength of Concrete with Compressive Strength of Cylinders and Cores, Pulse Velocity and Rebound Hammer", discussion, *ACI Journal* **78**, 152–155, Figures A and B.

16. Stamenkovic, H. (1979) "Short Term Deflection of Beams", discussion, *Journal of ACI* No. 2, Proceedings **76**, pp. 370–373, Figure Ba- Forces A and B.
17. Amerhain, J. E. editor, *Masonry Design Manual*, Masonry Industry Advanced Committee, 1971, 2nd edn, p. VII-46, Figure 12; 1979 edn, p. 89, Figure 222.
18. Marti, P. (1985) "Basic Tools of Reinforced Concrete Beam Design", *ACI Journal*, Proceedings **82**, 47, Figure 1a and b.
19. Saliger, R. (1949) *Prakticna Statika (Praktische Statik)*, Wein 1944, Yugoslav translation Nakladni zavod Hrvatske, Zagreb, Yugoslavia, p. 176, Figure 209 and 210.
20. Timoshenko, S. and McCollough, G. H. (1949) *Elements of Strength of Materials*, 3rd edn, 6th printing, D. Van Nostrand Company, New York, pp. 93–96, Figure 11; p. 137, Figure 159; pp. 94, 95; p. 148, Figure 172b; p. 72.
21. Timoshenko, S. and Young, D. (1968) *Elements of Strength of Materials*, 5th edn, D. Van Nostrand Company, New York, p. 62, Figure 3.11; pp. 28, 29, 194.
22. Wang, C. and Salmon, C. (1965) *Reinforced Concrete Design*, International Textbook Company, Scranton, Pennsylvania, p. 63.
23. Winter, G. and Nilson, A. H. (1973) *Design of Concrete Structures*, 8th edn, McGraw-Hill Book Company, New York, p. 62, Figure 2.13 and 2.14.
24. Sears, F. W. and Zemansky, M. W. (1953) *University Physics*, Addison-Wesley Publishing Company, Cambridge, MA, p. 33, paragraph 3-3.
25. Seely, F. B. and Smith, J. O. (1956) *Resistance of Materials*, 4th edn, John Wiley and Sons, New York, pp. 125–128, Figure 128.
26. McCormac, J. C. (1975) *Structural Analysis*, Harper and Row, New York, pp. 15, 22.
27. Freeman, S. (1974) "Properties of Materials for Reinforced Concrete", *Handbook of Concrete Engineering*, ed M. Fintel, Van Nostrand Reinhold Company, New York, p. 154.
28. Krauthammer, T. "Analysis for Shear Effects in Reinforced Concrete Beams and Slabs", *Developments in Design for Shear and Torsion*, Symposium paper, Annual Convention, Phoenix, AZ March 8, 1984, p. 66, Figure 2.
29. Borg, S. F. (1983) *Earthquake Engineering*, John Wiley and Sons, New York, pp. 88–89, Figures 7.6 and 7.7, Table 7.1.
30. Cumming, A. E. and Hart, L. (1953) "Soil Mechanics and Foundation", *Civil Engineering Handbook*, 4th edn, editor in chief Leonard Church Urquhart, McGraw-Hill Book Company, New York, pp. 8–10, Figure 5.
31. Kotsovs, M. D. (1983) "Mechanisms of Shear's Failure", *Magazine of Concrete Research* **35** (123).
32. Corley, W. G. (1966) "Rational Capacity of Reinforced Concrete Beams", *Journal of the Structural Division*, ASCE **92**, p. 131, Figure 5a – Beam N5.
33. Popov, E. P. (1976) *Mechanics of Materials*, 2nd edn, Prentice-Hall, Englewood Cliffs, New Jersey, p. 180, Figures 6.14b, g.
34. Hsu, T. T. C. (1982) "Is the 'Staggering Concept' of Shear Design Safe?", *ACI Journal* **79**, 435.
35. Mirza, J. A. and Furlong, R. W. (1985) "Design of Reinforced and Prestressed Concrete Inverted T Beam for Bridges", *PCI Journal* **25**, 112–136.
36. Stamenkovic, H. "Quantitative Evaluation of Shear Strength in a Flexural Member", *Development in Design for Shear and Torsion*, Annual Convention of the ACI, Phoenix, AZ, March 8, 1984, pp. 105–109, Figure 3.

37. Kommers, J. B. (1959) "Mechanics of Materials", *Civil Engineering Handbook*, 4th edn, editor-in-chief Leonard Church Urquhart, McGraw-Hill Book Company, New York, pp. 3–34, Figure 43.
38. Bassin–Brodsky–Wolkoff (1969) *Statics Strength of Materials*, 2nd edn, McGraw-Hill Book Company, New York, p. 250, Figure. 11.4, 11.5.

Chapter 3

Practical applications of the new theory

The fallacy of the truss analogy theory for reinforced concrete beams

3.1 A brief overview of the problem

This chapter shows that the truss analogy theory, as suggested by Ritter (1899)[1] and supported by Morsch (1902 and 1906),[2] appears to be incorrect. It demonstrates how in a flexurally bent reinforced concrete (RC) beam (either cracked or uncracked) there cannot be any truss formation, and that there is no similarity between a cracked RC beam and a real truss in the behavior of load distribution and stress condition. As long as the neutral axis is present, flexural tension and compressed struts (between any two adjacent stirrups) cannot coexist at the same time and at the same place. If a diagonal crack passes through the entire compressed zone, then an arch, with its hinge at the compressed zone, is created but not a truss.

A close examination and thorough study of five of Ritter's and Morsch's concepts (horizontal compressed strut, two compressed struts at supports, compressed struts between any adjacent two web stirrups, tensile chords of vertical stirrups and horizontal tensile chord of tensile reinforcement) reveal that each one of them is erroneous, as is the result of the efforts of those early pioneers trying to control, and eventually prevent, diagonal failure of RC beams.

It is true that tensile reinforcement is indeed in tension, but this fact alone does not prove the existence of formation of a truss in a cracked RC beam because elimination of the neutral axis will create a pure arch but never a truss.

Above all, it is emphasized that any equation involving the truss analogy theory was wrongly applied in attempts to solve the diagonal cracking problem, since the direction and magnitude of the resisting shear force V_r has been used in lieu of the shear force V. Therefore, it was impossible to ever solve a diagonal cracking problem in a RC beam.

Furthermore, there is a theoretical explanation that diagonal cracking in an RC beam is caused by the resultant punch shear force V_n. Punch shear force V_n is created by a combination of vertical shear force V_1 (caused by the support) and vertical shear force V_2 (caused by the external load) with flexural tensile force T. This is clearly illustrated in Figure 3.1. Yet, such an explanation is fundamental proof of the fallacy of the diagonal tension theory.

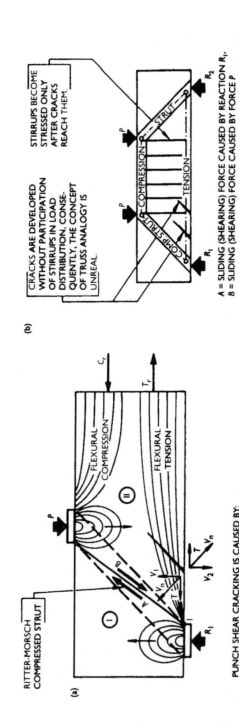

RITTER-MORSCH COMPRESSED STRUT

(a)

CRACKS ARE DEVELOPED WITHOUT PARTICIPATION OF STIRRUPS IN LOAD DISTRIBUTION. CONSEQUENTLY, THE CONCEPT OF TRUSS ANALOGY IS UNREAL.

STIRRUPS BECOME STRESSED ONLY AFTER CRACKS REACH THEM.

(b)

A = SLIDING (SHEARING) FORCE CAUSED BY REACTION R_1,
B = SLIDING (SHEARING) FORCE CAUSED BY FORCE P

PUNCH SHEAR CRACKING IS CAUSED BY:

V_1 = VERTICAL SHEAR FORCE CAUSED BY SUPPORT R_1, V_2 = VERTICAL SHEAR FORCE CAUSED BY EXTERNAL FORCE P, T = FLEXIBLE TENSILE FORCES,
V_n = RESULTANT PUNCH SHEAR FORCES.

FIG. a SHOWS OPPOSITELY ORIENTED COMPRESSIVE STRESS DISTRIBUTION CAUSED BY EXTERIOR LOAD P AND REACTION R_1, A FLEXURALLY COMPRESSIVE AND TENSILE STRESS DISTRIBUTION IS ALSO SHOWN. FORCE R_1 HAS TENDENCY TO SLIDE PORTION I UPWARD, WHILE FORCE P HAS A TENDENCY TO SLIDE PORTION II DOWNWARD. THE SHEAR–SLIDING PLANE IS LOCATED PRECISELY IN THE FIELD OF THE RITTER–MORSCH "COMPRESSED" STRUT. EVIDENTLY THIS TYPE OF STRUT CANNOT SERVE AS A COMPRESSION CHORD OF THE TRUSS BECAUSE OF THE NATURAL TENDENCY TO SLIDE IN TWO OPPOSITE DIRECTIONS.

FIG. b SHOWS TRUSS ANALOGY AS PROPOSED BY RITTER (1899) AND SUPPORTED BY MORSCH (1902). THE CRACKED R.C. BEAM DOES NOT ACT OR REACT AS A TRUSS BECAUSE DIAGONAL CRACKING IS DEVELOPED WITHOUT ANY REACTION OF THE STIRRUPS AS TRUSS ELEMENTS; ONLY THE STIRRUPS CROSSED BY CRACKS BECOME STRESSED. ALL OTHERS REMAIN IDLE.

(c) FIG. c SHOWS THAT A COMPRESSED STRUT CAN NEVER PENETRATE INTO THE TENSION ZONE OF AN R.C. BEAM: FLEXURAL TENSION AND COMPRESSION IN A GIVEN R.C. BEAM CANNOT EXIST AT THE SAME TIME AND AT THE SAME LOCATION. THIS FACT BY ITSELF PROVES THE TOTAL FALLACY OF THE TRUSS ANALOGY THEORY.

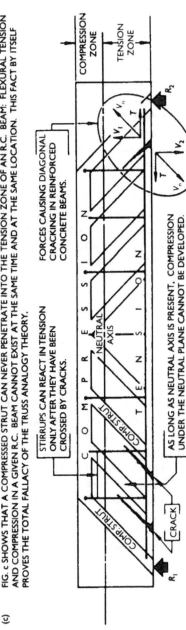

FORCES CAUSING DIAGONAL CRACKING IN REINFORCED CONCRETE BEAMS.

STIRRUPS CAN REACT IN TENSION ONLY AFTER THEY HAVE BEEN CROSSED BY CRACKS.

AS LONG AS NEUTRAL AXIS IS PRESENT, COMPRESSION UNDER THE NEUTRAL PLANE CANNOT BE DEVELOPED.

NOTES

1. THE OCCURANCE OF ONE OR MORE CRACKS CANNOT CAUSE ALL STIRRUPS TO BE IN TENSION AT THE SAME TIME, CONSEQUENTLY NO RIGIDITY OF ONE TRUSS COULD EVER BE CREATED;

2. FORMATION OF THE CRACK DOES NOT ELIMINATE THE NEUTRAL PLANE; IT IS ONLY MOVED UPWARDLY BUT IT IS STILL THERE;

3. POSSIBLE ELIMINATION OF THE NEUTRAL AXIS CAN BE ACHIEVED BY PRESTRESSING A BEAM BUT IN THAT CASE AN ARCH IS FORMED, NOT A TRUSS;

4. PRESENCE OF THE NEUTRAL AXIS STIPULATES THE EXISTANCE OF FLEXURAL TENSION WHICH, IN TURN, EXCLUDES ANY POSSIBLE PENETRATION OF A COMPRESSIVE STRUT INTO THE TENSILE ZONE;

5. AS LONG AS THE NEUTRAL AXIS IS PRESENT, A TRUSS CAN NEVER BE CREATED IN AN R.C. BEAM.

Figure 3.1 This figure shows that the mechanism of truss analogy, as suggested by Ritter and Morsch, could never be developed to react as a real truss in a cracked or uncracked RC beam.

This work is supported by published materials, the laws of physics, the laws of statics, common sense and original evaluation.

3.2   Introduction

The design procedure for an RC beam in shear has been based on the truss analogy theory which was developed at the turn of the century by Ritter[1] in 1899 and by Morsch[2] in 1902. "It has formed the principal basis for the interpretation of forces in beams and for the design of RC beams for shear."[3] This theory postulates that a cracked RC beam acts as a truss formed by parallel longitudinal chords and web chords (composed of diagonal concrete struts, penetrating into a flexural tensile zone of the concrete beam and touching the tensile reinforcement) and transverse steel stirrups as tensile chords.

Test data supported this assumption because diagonal cracking can indeed be stopped by vertical stirrups. But Ritter and Morsch failed to note that such diagonal cracking can also be caused by another phenomenon called "punch shear"[4,5] (as shown in Figures 3.1(a) and (c)) and that such cracking can be stopped by vertical reinforcement (adequately anchored in a concrete mass to prevent its own bond failure) across the diagonal cracking. Such reinforcement does not need to touch or be connected to horizontally compressed "chords." For example; stirrups, located as vertical reinforcement only in the tensile zone and as a completely independent reinforcing element from the horizontally compressed strut, will also prevent diagonal cracking in a deeper concrete beam if such cracking strikes the stirrups in their lower portion, provided there is adequate anchorage above. This independent web reinforcement cannot and does not represent any truss member.

The dissimilarity between the truss analogy theory and reality can be inferred from the following known facts: "Experience with the 45-degree truss analogy revealed that the results of this theory were typically quite conservative, particularly for beams with small amounts of web reinforcement."[6] In 1922, Morsch made the following statement: "We have to comment with regards to practical application that it is absolutely impossible to mathematically determine the slope of the secondary inclined cracks according to which one can design the stirrups."[7] Further doubts about the truss analogy explanation for a RC beam are voiced in a report of the Joint Committee ASCE and ACI: "The observed behavior of a beam with web reinforcement is more complicated than is indicated by the truss analogy on which the current design procedures are based."[8] . . . "In the absence of detailed knowledge regarding the stress distribution for both shear and flexure, a fully rational design approach to the problem does not seem possible at this time," because "distribution of shear and flexural stress over a cross section of RC are not known."[8] Furthermore, as can be seen from the published discussion,[9] leading authorities on shear and diagonal tension have been unable to negate any of the six powerful statements made by us concerning the fallacy of the truss analogy theory. This leads to the conclusion that the truss analogy theory was a contrived concept that cannot withstand even the slightest challenge, as can be seen more clearly in the following discussion in which the dissimilarity between a cracked RC beam and the truss analogy theory will be shown clearly.

3.3 Theoretical considerations

In his comments in the paper "Quantitative evaluation of shear strength in flexural member,"[10] Professor Gordon Batson (Clarkson University, Potsdam, NY) made the following statements: "I believe the main points of the first few pages is that there is a difference between resisting shear forces (stresses) and active shear forces (stresses) and that one is used for equilibrium purposes and the other for computing stresses." This is the basis for understanding the fallacy of diagonal tension. In his last statement he said: "In summary you make a strong argument for the distinction between the active shear forces and resisting shear forces. This point was strongly emphasized in old strength-of-materials textbooks, but more recent textbooks simply use equilibrium of the free body for a portion of a beam in bending, and thus are not concerned with $V = V_r$ at any section." This means that on the face of a portion of a free body diagram, vertical shear forces V and their corresponding vertical resisting shear forces V_r, of equal magnitude but opposite orientation, do exist and act simultaneously. Also, there exist and act simultaneously compression (C) and tensile (T) forces and their balancing resisting compression (C_r) and resisting tensile force (T_r), of which the last two (T_r and C_r) equilibrate a portion of the free body diagram. Naturally, an action and its reaction (or active and resisting forces) are located in the same plane, along the same line and are of equal magnitude but opposite orientation, as shown in Figure 3 of reference 11. Furthermore, this means that resisting vertical shear forces V_r equilibrate an algebraic sum of external-active forces (reaction R and bending forces F) while vertical shear force V equilibrates an algebraic sum of resisting reaction forces (R_r) and resisting bending forces (F_r), as is illustrated by Figures 2(a) and 2(b) of reference 11.

Yet, what Professor Batson saw and recognized has not been envisioned by classical theoreticians, who treated vertical resisting shear force V_r (at the face of a portion of the free body diagram) as an active force V. Such an error in treating the application of *resisting* (compression, tensile, vertical and horizontal shear) forces on a free body and treating them as active (compression, tensile, vertical and horizontal shear) forces, has been so deeply rooted in the minds of educators that resisting compression force C_r (applied for equilibrium of a portion of the free body diagram) has been used erroneously in the same location as active compression force C. In that sense, Professor Kani[12] literally combined (in his Figure 5(a)) compression resisting force C_r with external concentrated load P and shows a "resultant" of this combination. Evidently, everyone knew that such resisting compression and tensile forces could be replaced by resisting bending moment M_r (for a given cross section), as is illustrated in reference 13, Figure 128(b). Yet, when one came to the explanation of diagonal cracking in flexural bending, the reality of the resisting moment M_r, which could be created only by resisting forces (C_r and T_r), has been lost. Clearly, no attention was paid to the error in this or any of the other cases dispersed throughout structural engineering literature). The fact is that the error of combining active bending forces with resisting forces (or combining actions and reactions) is intolerable in engineering mechanics. But, because no one knew what was wrong and what was right concerning diagonal cracking in a flexural member, the stated solutions have been accepted as the correct and even used up to the present. For that reason, it should not be surprising that the concept of diagonal cracking in a flexural member has not been understood. How could any of us comprehend the concept of diagonal tension cracking when we applied resisting shear force V_r for equilibrium of a free body diagram and yet treated such a force as an active force V, believing that such a force indeed causes diagonal tension. Consequently, if one applied resisting shear force V_r in the equations as active force V, in a direction opposite to the action to active force V, how could diagonal cracking correspond to its bending forces?: "The test results obtained from several hundred beams have shown that no direct relation between 'shear strength' and shear force V exist,"[14] ... "However, with the development of cracks, an extremely complex pattern of stresses ensues and many equations currently in use have little relevance to the actual behavior of the member at this stage."[15]

In order to find a solution and describe diagonal cracking analytically, Ritter[1] and Morsch[2] suggested a truss model to simulate the reaction of an RC beam subjected to flexural bending. After diagonal cracking in the RC beam, the concrete was supposed to be separated into a series of diagonal concrete struts (Figure 3.1). Such concrete struts should interact with the stirrups at their top, and the tensile reinforcement (forming a plane truss), capable of resisting the imposed bending forces as well as shear forces; even though "the fourth assumption of the truss analogy, that the diagonal tension crack forms up to a vertical height j_d, is an assumption of convenience without support from laboratory test results."[8]

The compression zone of the concrete and the tensile reinforcement are supposed to serve as the top and bottom chords of a truss. The stirrups are treated as the tensile web member while the portion of concrete between any two adjacent cracks is treated as a compression web member, even though that compression strut never could never develop in a tensile zone as long as the neutral axis in a bent member exists! (Elimination of the neutral axis occurs in only two ways: one, by collapsing of a member; and two, if a crack reaches the top level of the compressed zone and the beam still does not collapse. Then the beam will never react as a truss, but rather as an arch with its hinge in the compressed zone.) So it has been internationally accepted that the impossible concept is possible even though "It has been pointed out in this report that the classical procedures are questionable in their development, as well as misleading and sometimes unsafe in their application,"[7] ... "In other words, the classical design procedure does not correlate well with the test results."[8]

The additional irony of the truss analogy theory is that Ritter[1] and Morsch[2] assumed that the angle of inclination of diagonal cracking was 45° even though: "Failure occurs with a sudden extension of the crack toward the point of loading,"[16] ... "Under a single point load, diagonal failure occurs closer to applied load rather than to the support," Figure 1(a),[17] which fundamentally negates the angle of 45° or any other angle, as assumed by the truss analogy theory.

As late as 1969, Lampert and Thurlimann[18] assumed that the angle of inclination of diagonal struts was variable, "depending on the volume ratio of longitudinal steel into transversal steel."[19] The angle of cracking was been theoretically established ("This angle is constant throughout the beam, except at support and midspan where the local effect known as the 'fanning' of the concrete struts occurs,")[19] even though engineers know that the angle is not constant, "Figure 7.16 illustrates that at the interior support of a beam, the diagonal cracks, *instead of being parallel*, tend to radiate from the compression zone at the load point."[15]

Because of the lack of any present knowledge or understanding of diagonal cracking in a flexural member, many new expressions have been introduced in literature, one of which belongs to the truss analogy theory and is known as the "fanning" phenomenon. In fact, the "fanning" phenomenon is natural cracking caused by the moving external load toward the support, which is partially discussed in Section 3.2 of this chapter.

We now try to resolve the above misunderstanding, with the help of the following discussion.

3.4 Discussion

As stated earlier in the introduction and in the discussion of theoretical considerations, the hypothesis of truss analogy is incorrect because no similarity or resemblance in form or behavior exists between a cracked RC beam and any truss. This can be seen as follows.

3.4.1 Upon formation of cracks, the flexural compression zone of a reinforced concrete beam cannot be converted into an axially loaded strut

The essence of a chord's loading (as a member of any truss) is axial loading, where flexural deflection of the chord's weight is neglected. In a flexurally bent beam, the entire load is imposed in bending only. When some cracks develop, the neutral axis moves up and the compressed cross section decreases, but deflection increases rapidly as an element subject to flexural bending must do. Formation of one or more cracks (in an unfailed beam) can never cause the flexural compressed zone of an RC beam to be eliminated and replaced by axial loading (in order to be treated as a compressed chord) for the following reasons:

1   In the compression zone, the deflection that has already occurred, due to concentrated loads (Figure 3.1(c)), must be removed in order for that zone to act as a straight compressed strut of a truss.

2   To convert the existing flexural compression into axial compression and to change the bending curve into a straight line of a chord, vertical bending forces must become horizontal forces acting axially through the compressed chord.

3   Such a change from flexural bending loading to axial compressive loading is possible only if new axial forces are imposed.

4   By eventual replacement of bending forces with axial forces (pure bending), transversal shear forces are eliminated. The question arises: How can we calculate and control the diagonal cracking in such a member (strut) in which transversal shear forces do not exist?

5   As long as the flexural compression of an RC beam exists as an unbroken unit (independent of how shallow it becomes) and flexural tensile reinforcement still acts in tension (even though some cracks are already formed in the tensile zone), the neutral axis of such a beam is still there, located at the bottom of the flexural compressed zone (now decreased due to cracking).

6   Yet, as long as the neutral axis exists, it is not possible for any compressed strut, as a member of a truss, to be formed in such an RC beam.

7   Also, possible elimination of the neutral axis in an RC beam could be achieved by the development of a crack through the entire compressed zone. However, in that case, failure occurs. If failure does not occur, an arch with its hinge at the top will be created, but never a truss.

8   Even if two, in slope, compressed struts above the supports are formed (which can never happen, as illustrated in Figures 3.1(a) and (b)), the already existing flexural stresses in the "horizontal strut" cannot be released in order to be replaced by axial compressive stresses.

9   Based on the above facts, it is clear that Ritter and Morsch made an obvious error by suggesting that flexural compression in an RC beam, upon formation of some cracks, can ever react as a compressed strut of a given truss.   .

3.4.2 Compressed struts above supports cannot be developed in a reinforced concrete beam

For a better understanding of the fallacy of the truss analogy theory, it is imperative to understand and accept the fact that the natural tendency of any physical support is to punch out (to slide upward) some portion of the deformed beam to release itself from its compressed condition. Also, the natural tendency of an external load (force) is to slide (to punch downward) its portion of a beam, for this portion to be freed from its deformed position and stressed condition. Compressive stress distribution of the support is oriented upward while compressive stress distribution of a nearby external load is oriented downward. Such stress distributions can have some matching or borderline location (depending on the distance between the force of the external load and the force of the support reaction) which coincides with the border of the so-called struts of the truss.

As shown in Figure 3.1, under the action of the external force R_1, the portion of the beam denoted by I has a natural tendency to move upward. The portion of the beam denoted by II has a natural tendency to slide downward, as a result of the action of the external force P. Since portions I and II have a tendency to move in opposite directions, a compression strut (as Ritter and Morsch suggested) cannot be developed in the RC beam.[20] On the contrary in the field designated as "strut," a sliding–shearing plane is developed with shearing forces A and B. In fact, through this "strut," diagonal cracking usually occurs starting at point 1 precisely as any pullout test proves: the crack is a straight line between the pullout force (here, concentrated load P – Figure 3.1(a)) and the counter-pressure force (here, reaction R_1 – Figure 3.1(a)). In other words, such diagonal cracking is caused by forces such as those illustrated in Figures 3.1(a) and (c):

1   Vertical shear force V_1, caused exclusively by the support and oriented as the support itself – here, upward.
2   Vertical shear force V_2, caused exclusively by the external load (uniform, concentrated, or a combination of both) and oriented as the load itself – here downward.
3   Flexurally tensile force T, caused by bending forces.
4   Resultant punch shear force V_n, caused by a combination of vertical shear forces V_1 and V_2 with horizontal tensile forces T, acting perpendicularly from any crack.

    Obviously, the resultant punch shear force V_n will be created for any type of load: concentrated, uniform, etc. As long as pure bending does not exist (where transversal forces V_1 and V_2 are eliminated), the resultant punch shear force V_n will always be present, as a result of the combination of the vertical shear forces (V_1 and V_2) with flexural tensile forces (T), leading to diagonal cracking of a bent member.

3.4.2.1 Any flexural crack proves existence of internal active and internal resisting forces

The natural tendency of a support to slide its portion (I) (Figure 3.1(a)) upward due to shown sliding force A, and the tendency of the external load to slide its portion (II) (Figure 3.1(a)) downward due to sliding force B could be explained as follows:

•   A combination of shear force V_1' (caused by the support) with corresponding compression force C (in the compressed zone) creates a resultant sliding force A which acts parallel to the cracking (here, upward), as shown in Figures 3.2(a) and (b).
•   The combination of shear forces V_2' (caused by the external load) with corresponding compression force C creates a resultant sliding force B, which also acts parallel to cracking but here downward, as is illustrated by Figures 3.2(a) and (b).

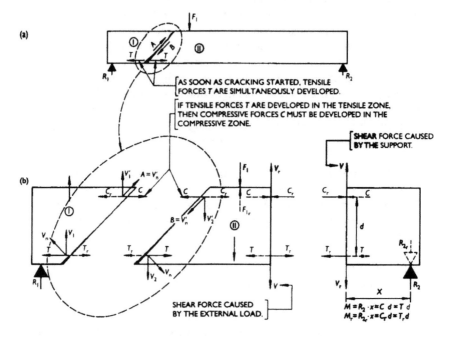

Figure 3.2 This figure graphically illustrates that the combination of internal active compressive and tensile forces with internal active vertical shear forces causes diagonal cracking and diagonal failure.

For direction (orientation) of tensile forces T at the cracking area, there is no dispute in the literature: they must act in directions opposite to each other (left and right) from the crack. However, to cause such cracking, these forces must be internal active forces T – here too, there is no dispute. Therefore, in order for the internal tensile force T to be in equilibrium on a free body (I) at the left-hand side, there must coexist a corresponding internal active compression force C on the same free body (I), acting in an opposite direction to the tensile force T, as shown in portion (I) of Figure 3.2(b). In other words, if internal active tensile force T exists (as it evidently does), then internal active compression force C must coexist because in flexural bending, there must simultaneously coexist flexural tensile and flexural compression zones. So, by observing only one flexural crack, the existence of internal active tensile and compression forces has been established by statical means.

Moreover, looking at the left portion (I) as a free body, and comparing the new and old force C (shown as C_r in Figure 3.2(b)), it becomes evident that they are collinear but oppositely oriented, and that the existing force C has always been used for equilibrium of a free body. It is the same with the tensile force T, which is collinear and oppositely oriented to the old tensile force T, which has been used for equilibrium of a free body. In other words, for any flexural crack to occur, two internal active forces T must exist, acting in directions opposite to each other, as shown in Figure 3.2(a). Due to the flexural compression and tensile zones, if internal active tensile force T exists, then internal active compression force C must coexist, as shown in Figure 3.2(b). However, if such forces are active (as they are), the forces shown in the cross sections in any textbook must be internal resisting compression C_r and tensile T_r forces, as shown in Figure 3.2(b).

With such logic, we satisfy Sears–Zemansky's statement that "a simple, isolated force is therefore an impossibility",[21] and Timoshenko–Young's statement that "to have equilibrium of a body, it is not necessary that the active forces alone be in equilibrium but that the active forces and reactive forces together represent a system of forces in equilibrium".[22] Also, we satisfy the second principle of engineering mechanics for equilibrium: "Two forces can be in equilibrium only in the case where they are equal in magnitude, opposite in direction, and collinear in action."[22] This principle has never been satisfied by the classical theory because, even though the compression force C is in translational equilibrium with tensile force T and the upper horizontal shear force H (Figure 5.1) is in translational equilibrium with the lower horizontal shear force H, they are not in rotational equilibrium, even though $\sum X = 0$. This is because only three moments (couples) exist in the longitudinal vertical plane of the bent member. Couple T–C could be equilibrated by couple V–R_1, but couple H–H cannot be equilibrated by any other couple because there is no other couple available. Yet, this is probably the main reason that Jack McCormac concluded in his textbook, "These equations $\left(\sum X = 0, \sum Y = 0 \text{ and } \sum M = 0\right)$ cannot be proven algebraically."[23]

Finally, the above explanation of diagonal cracking of a concrete member offers the real and only answer to McCormac's conclusion that "Despite all this work and all the resulting theories, no one has been able to provide a clear explanation of the failure mechanism involved."[24]

Consequently, with the above investigation of internal tensile forces T which participate in causing diagonal cracking in a flexural concrete member, it appears evident that besides internal active tensile force T, internal active compression force C exists, too! Also, it appears evident that such active internal forces T and C are oppositely oriented to the internal resisting compression and tensile forces which have been used by textbooks to establish equilibrium of a free body.

Evidently, simple cracking in the tensile zone of a concrete member leads us to discover and statically prove the existence of internal active and internal resisting forces in any flexurally bent member. This is in full agreement with analytical proof and calculation of the same forces as shown in reference 11.

The angle of cracking is a highly variable parameter. For a concentrated load at some distance from a support, it will be at approximately 45° from the load at the left or right side. For a concentrated load nearer to the support, it will be a straight line between the support and the concentrated load. For a uniform load, it will be at approximately 45° from the support, upward. For a deep beam, it will be a straight line between the external load and the support, while for a uniform load it will be a straight line between the support and approximately one-third of the length at the top of the beam. As a result of arch creation, cracking is prevented from reaching the middle top of the span. But the angle of cracking is totally irrelevant. Diagonal cracking is caused by punch shear forces V_n and not by any diagonal tension, as is seen clearly in Figures 3.1(a) and (c). So by proving the fallacy of the truss analogy theory, the fallacy of the Ritter–Morsch concept of diagonal tension is simultaneously proved, which Figures 3.1(a) and (c) indisputably demonstrate.

Ritter's concept that a compressed strut is developed between the support and the exterior load is unfounded, since it is based upon an incorrect assumption, as has been clearly proved by us elsewhere[4] and is easily seen in Figure 3.1(a), which shows that a strut is not created. A shearing plane of two oppositely oriented stress distributions, caused by the support (upward) and by the external load (downward), is present. A look at Figure 3.1(a) prompts the question: How can a compression strut be developed where a natural tendency exists for two portions of the beam to move in opposite directions? A compressed strut between support and concentrated load in a flexurally bent beam cannot be developed or formed in the same way, as in a pullout test a compressed strut cannot be developed between pullout force and counterpressure force.

It can be said that if a vertical exterior force P can create a compressed strut on a 45° angle, by the same analogy it follows that the physical support R_1 can create its own compressed strut. Evidently, these two oppositely oriented struts will have oppositely oriented compressive stress distributions with opposite tendencies for movement (sliding). Consequently, between these two struts, a sliding (shearing) plane will develop, as illustrated in Figure 3.1(a). Therefore, a single strut cannot be developed as part of a truss in an RC beam[20] as Ritter–Morsch believed it could.

To support the suitability of the truss analogy theory for an RC beam, a symposium paper in Phoenix[25] has demonstrated the failure of a deep beam caused by concentrated loads. Without any exception, the direction of failure appears to be a straight line between any support and a corresponding load. When this author wanted to know the location of a compressed strut between any support and corresponding load, the symposium speaker, naturally, could not answer. Instead, he showed one stripe developed *outside of the support* (leading toward the concentrated load) as a result of the so-called "fanning" phenomenon. All cracks shown to cause failure of a deep continuous beam, appear to represent pure pullout test failures on a much larger scale, without the slightest sign of any truss struts or any type of diagonal tension failure.

From the foregoing discussion, it appears to be impossible to ever develop a real compressed strut between a support and the exterior load in an RC beam unless special prestressing between the exterior load and support is made. Finally, in a pullout test, how could a compressed strut ever be developed between pullout force (P) and counter-pressure force (R_1) (Figure 3.1(a))? Evidently, a rational answer does not exist for such a question, as Figure 3.1(a) illustrates.

3.4.3 Web reinforcement (as tensile members of the truss) does not participate in load distribution in a cracked or uncracked reinforced concrete beam

Through any pin of a real truss, an external load will be proportionally distributed in any chord. *During loading of a truss, if the load is not distributed in a particular chord, that chord is not a static part of such a truss.* Any vertical stirrup of the Ritter–Morsch's truss becomes loaded only after a crack has developed in the deformed RC beam and crossed such a stirrup. *This shows that the beam never reacted as a truss and only the stirrup intersected by diagonal cracking is loaded in tension.* All other stirrups remain idle, free of tension, even though the beam is near collapse.

In order to support the above facts, let us quote from some authoritative publications:

> "When diagonal cracking occurs, web bars intersected by a crack immediately receive sudden increases in tensile stress in the vicinity of the crack, while web bars not intersected by diagonal cracks remain unaffected."[8]

> "The web reinforcement being ineffective in the uncracked beam, the magnitude of the shear force or stress which causes cracking to occur *is the same as in a beam without web reinforcement*."[26]

> "The largest strains measured occurred in each stirrup where it was crossed by the potential failure crack. It is seen that during the first five cycles of loading some 40 kips were resisted by mechanism not involving stirrups."[15]

> "Stirrups cannot be counted on to resist shear if they are not crossed by the inclined crack."[3]

> "Although web reinforcement increases the ultimate shear strength of a member, it contributes very little to the shear resistance prior to the formation of the diagonal tension cracks."[27]

Permanent small stresses in the stirrups can be understood in terms of Poisson's ratio.

So, vertical or inclined web stirrups acting in tension are disregarded because this reinforcement remains idle until propagation of already developed cracking in the concrete reaches these stirrups.[4,15,26,28] The loaded RC beam does not act or react as a truss because diagonal cracking in the RC beam is developed without any reaction of the stirrups as truss elements. Even a cracked beam can never react as a truss.

In Figure 3.3, the configuration shown is not rigid without the participation of all stirrups and, therefore, cannot be considered a truss. The participation of one or more stirrups in tension, as the result of the intersection by cracks, does not prove rigidity; the uncracked concrete between any two stirrups is a rectangular configuration without any rigidity. The term rigid, as used here, means that a truss, as a geometrical configuration, must not collapse, as shown by the dotted line in

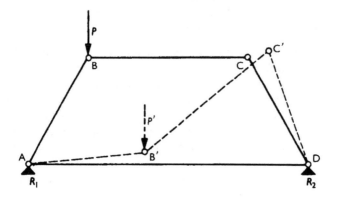

Figure 3.3 Lack of rigidity may result in failure under loading.

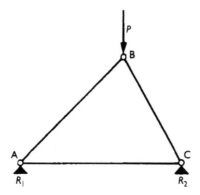

Figure 3.4 Rigidity of a simple triangular truss.

Figure 3.3. Figures with four or more open sides are not stable and may collapse under load. Figure 3.4 shows the essential concept of the truss theory: *rigidity*. The rigidity of a simple triangular truss is self-explanatory. A beam where cracking has already developed and, thus, has caused one or more of the stirrups to be in tension (but not all of them) is not a rigid configuration and cannot be treated as a truss, so the truss analogy theory cannot be applied. Rigidity is the essence of any truss, so any rectangular configuration, formed by a couple of stirrups which are not in tension, will destroy the truss analogy.

It is evident that the Ritter–Morsch's truss analogy theory cannot be applied to an RC beam because such a truss cannot exist if stirrups are not a static part of the assumed truss.

The assumption that pins B and C in Figure 3.3 can be imagined to be rigid due to the presence of surrounding concrete is unacceptable because special horizontal and vertical forces will be developed immediately after imposing load P_1. Also, by assumption of existing rigidity of pins B and C, we enter into the field of rigid arches and abandon the truss concept. The theoretical differences between two such static members are enormous. The fallacy of creating an arch and its contribution to diagonal cracking in an RC beam is explained by us elsewhere.[4] Therefore, because the assumption of a static element known as an arch for the truss concept is false, it follows that any conclusion from such a theory must be erroneous.

3.4.4 A compressed strut between two stirrups cannot be developed

The basic assumption of the truss analogy theory, that struts are formed between any two adjacent stirrups immediately after some cracks have developed in a RC beam, is fundamentally wrong and unacceptable for the following reasons:

1   Before the crack reaches a stirrup, that stirrup has been idle – free of any tension.[15,26,28]

2   When the crack reaches the stirrup, the neutral axis is moved upward, but it is still there, with an enlarged flexural tensile zone, where a compressed strut cannot be present.

3   Tension caused by the intersection of a stirrup with a crack does not cause tension in any other stirrups, so the essential law of a truss, known as the law of rigidity, is not satisfied. Consequently, the mechanism of the truss cannot be made workable.

4   As long as the neutral axis is present in an RC beam, it is a ridiculous idea that a compressed strut can ever be developed in a flexural tension zone. Evidently, such an assumption does not seem to be possible.

5   During the collapsing procedure, if the truss mechanism has to develop in a fraction of a second, each stirrup must be struck simultaneously by some crack. This is practically impossible.

6   The statement from textbooks (regarding the role of web reinforcement) that the concrete block between two adjacent cracks in a tensile zone acts as a compressed strut by "suppressing flexural tensile stresses in the cantilever blocks by means of the diagonal compression force Cd, resulting from truss action"[15] is not only wrong but also impossible because the laws of physics do not allow this to be so. How can there ever be suppressed flexural tensile stresses in a tensile zone where the neutral axis is present? Did anyone ever measure such a compression in a tensile zone? If "suppressed flexural tension" does not exist then the neutral axis must somehow be eliminated from the beam because nature does not allow such an abnormality.

7   Finally, if such concrete struts are indeed formed after the formation of a crack in the RC beam, then such compression in the flexural tensile zone must be evident to tensometers. Yet, this does not seem to have been reported – that a compressed strut and flexural tension could indeed coexist at the same spot

and at the same time; nor has the existence of any compression in the tensile zone of a flexurally bent beam ever been recorded. It is evident that such a concept as a concrete strut in a flexurally bent beam cannot be supported by common sense or any law of physics. This leads us to conclude that it may be fundamentally incorrect.

3.4.5 Without touching the compression zone, stirrups will prevent propagation of diagonal cracking

In a deep beam, stirrups could be located only in the tensile zone of the concrete beam and not extend beyond its neutral axis. Even with such a limitation, they will be capable of stopping further propagation of diagonal cracking if the cracks strike (touch) the lower portion of the stirrups. Naturally, the upper portions of the stirrups need to be anchored in the concrete mass only for a minimal length to be capable of preventing bond failure of the stirrups. The closer the spacing of stirrups in either case (stirrups covering the entire cross section and stirrups covering only the tensile portion of the cross section) the more the resistance to further propagation of already developed diagonal cracking. This is true because already developed diagonal cracks will strike (touch) a stirrup in its lower portion enabling the rest of the stirrup to serve as an anchorage in the concrete mass against pullout.

With such facts, how can it be said that stirrups can represent the tensile chords of a truss, even though they are separated from the compressed horizontal strut by the neutral plane of the beam itself? Neither common sense nor logic can support a statement that the prevention of further propagation of diagonal cracking by vertical reinforcement automatically concludes that such reinforcement must be part of a truss. On the contrary, such a fact can only lead to the conclusion that the truss analogy cannot be established for any configuration where chords are not connected to each other and stay as independent members at some distance from each other in the concrete mass. Further, the above discussion reveals that the truss analogy theory is not based on similarity or resemblance to some truss, but rather on an assumption which appears to be false.

3.4.6 The existence of a neutral axis excludes the existence of the truss concept

Formation of some diagonal cracks moves the neutral axis towards the top of the bent beam, but its presence is not eliminated in an unbroken beam that is partially cracked diagonally. The flexural tensile zone is still there; it only becomes deeper while the compression zone becomes shallower.

Because of binding forces (bond) between tensile reinforcement and concrete, some flexural tension must exist in the so-called "compressed strut" in the tensile zone. Flexural compression can never pass the neutral axis and penetrate into the tensile zone of the "compressed strut." It is only possible when the neutral axis starts to move into the flexural tensile zone and reaches the flexural reinforcement, and only then could a real compressed strut be developed. As long as the neutral axis is moving upward by propagation of cracks, such a strut can never be developed.

As stated earlier, the neutral axis and tensile zone are eliminated only when the crack reaches the top of the beam and the member still does not collapse, but then such a beam is converted into a pure arch with a hinge at its top. Yet, it is very interesting to note the statement in a textbook: "Truss mechanism in a beam can function only after the formation of diagonal cracks (that is, after the disappearance of diagonal tension) in the concrete"[15] but this overlooks the fact that a truss cannot function with the existing neutral axis of the beam, and the compressed strut of concrete in the tensile zone cannot be created. The possible exception for creation of a strut may be during the moment of the collapsing of the beam, which can be a fraction of a second. But, the existence of one (or two) sloping struts, (even for a fractional second) during the collapsing mechanism, does not cause or support the concept of "rigidity". This is because the elimination of the horizontal compressed zone (strut) and the existence of a rectangular shape between any two stirrups (without cracks) makes the configuration unstable, which negates the entire philosophy of truss analogy for a bent beam.

3.4.7 Horizontal tensile reinforcement does not determine the existence of the horizontal tensile chord of a truss

Because compression struts at supports can never be formed and because flexural compression indeed exists, tensile reinforcement represents tensile anchorage of an arch. Even if two struts were constructed by prestressing (between a support and a concentrated load), a truss could not function because a web strut can never be created to extend into a flexural tensile zone. Can any compressed strut exist in a flexural tensile zone? Yes, if special diagonal prestressing occurs there, it is possible to create such struts; but, without this special construction, diagonal, compressed struts cannot be formed by the occurrence of any crack in the tensile zone.

Also, tensile stirrups cannot represent vertical chords of a truss. The tension of stirrups is possible only for those stirrups which are crossed by the cracks in a concrete mass. Even if all stirrups are crossed by cracks, this does not eliminate the neutral axis of a reinforced beam. As long as a crack does not penetrate through the entire flexural compression zone, the neutral axis is present in an RC beam. The presence of the neutral axis makes it impossible for diagonal compressed struts to be created in the tensile zone. As soon as any crack penetrates through the entire compressed zone of an RC beam, the neutral axis will be eliminated, following eventual arch formation, with its hinge at the top of the cracked spot. So, possible formation of any truss cannot take place at the moment when the crack reaches the top of the flexural compressed zone. Consequently, in an RC beam, the *tension condition of the horizontal reinforcement could determine the existence of an arch but never of a truss.*

Thus, it can be seen that tensile reinforcement will never have a chance to serve as the tensile chord of a truss. The simple meaning of the above statement is that the concept of Ritter and Morsch is also erroneus, as are the other four concepts.

3.5 Final note

Any one of the five Ritter–Morsch concepts for acceptability of the truss analogy theory is enough to negate the hypotheses of the truss analogy theory. Seeing that none of their concepts is correct, it is evident that these two pioneers made a fundamental error in their assumption of possible truss creation in a cracked RC beam. *Consequently, it is obvious that the concept that the truss analogy theory can be used to calculate stresses, in order to control diagonal cracking in an RC beam, is not valid and should be abandoned, as a theory, forever.*

3.6 Conclusion

Based on the above discussion, the following conclusions can be drawn:

1 A horizontal compressed strut (in the compressed zone of a bent RC beam), axially loaded, can never be developed or formed except if a new axial load (such as prestressing) is imposed.
2 Two diagonal compressed struts, connecting supports and imposed load in an RC beam (as shown in Figure 3.1(a)) cannot be formed under any condition.[4]
3 An inclined compressed strut in a flexurally bent RC beam could be formed if special prestressing is applied between the support and the imposed load (or between tensile and compressed zones).
4 A compressed strut between any two stirrups cannot be formed in a flexural tensile zone as long as such a zone exists, irrespective of the number or types of diagonal cracks.
5 The formation of a crack or cracks in an RC beam does not eliminate the neutral axis; rather, they move the neutral axis upward, making a shallower compression zone.
6 The neutral axis could be eliminated by penetration of a crack through a compressed zone, but then a truss would not be formed, rather an arch with its hinge at the compressed zone.
7 As long as the neutral axis is present in an RC beam, flexural tension and the strut's compression (at the same spot in an RC beam) cannot coexist in order to form an inclined compressed strut between two stirrups.
8 Vertical web reinforcement is unstressed as long as it is not reached by cracks. When a beam is already cracked, only those stirrups reached by cracks in the concrete are loaded in tension while all other stirrups are idle.

9 One or more stirrups in tension (crossed by cracks) do not determine the concept of rigidity; consequently, to control the safety of an RC beam by analytical treatment as a truss is an engineering absurdity.

10 In order to control diagonal cracking, in any equation based on the truss analogy theory, the direction and magnitude of resisting shear forces V_r has been wrongly applied in lieu of shear force V: consequently, it was impossible to solve the diagonal cracking problem in an RC beam.

11 The entire Ritter–Morsch hypothesis concerning diagonal tension and the truss analogy theory appears to be wrong and should not be used for an engineering design of concrete structures.

12 Diagonal cracking in an RC beam is caused by a combination of vertical shear forces (caused by the support and the external load) and flexural tensile forces (as shown in Figure 3.2).

13 Determination of such forces, quantity and type of reinforcement have been explained in Chapter 2 of this book and, partially, in reference 11.

References

1. Ritter, W. (1899) "Die Bauweise Hennebique", *Schweizerische Bauzeitung*, Zurich, 33(7), 59–61.

2. Morsch, E. (1909) *Concrete-Steel Construction*, English translation by E. P. Goodrich, McGraw-Hill Book Company, New York, p. 368.

3. ACI-ASCE Task Committee 426 (1973) "The Shear Strength of Reinforced Concrete Members", *Journal of the Structural Division* ASCE 99, 1116–1146.

4. Stamenkovic, H. (1978) "Suggested Revision to ACI Building Code Clauses Dealing with Shear Friction and Shear in Deep Beams and Corbels", paper by J. G. MacGregor and H. M. Hawkins, discussion, *ACI Journal* 75, 222–224, Figure C and D.

5. Stamenkovic, H. (1977) "Inflection Points as Statical Supports are Responsible for Structural Failure of AMC Warehouse in Shelby, Ohio, 1955", *Materials and Constructions* 375–383.

6. Collins, M. P. and Mitchell, D. (1980) "Shear and Torsion Design of Prestressed and Non-Prestressed Concrete Beams", *PCI Journal* 25, 36.

7. Morsch, E. (1922) *Der Eisenbetonbau*, Verlag von Konrad Wittwer, Stuttgart, 128.

8. ACI-ASCE Committee 326, "Shear and Diagonal Tension", Proceedings, *ACI*, Vol. 59, January–February–March, 1962, pp. 3, 7, 18, 21, 51.

9. Stamenkovic, H. (1981) "Shear and Torsion Design of Prestressed and Non-Prestressed Concrete Beams", discussion, *Journal of PCI* 26(6), 106–107.

10. Stamenkovic, H., "Quantitative Evaluation of Shear Strength in a Flexural Member", *Developments in Design for Shear and Torsion*, Symposium paper, Annual Convention, Phoenix, AZ, March 8, 1984.

11. Stamenkovic, H. (1978) "Suggested Revision to Shear Provisions for Building Codes", discussion, *Journal of ACI* 75, 565–567.

12. Kani, G. N. J. (1964) "The Riddle of Shear Failure and Its Solution", *Journal of ACI* 61, 447, Figure 5a.

13. Seely, F. B. and Smith, J. O. (1956) *Resistance of Materials*, 4th edn, John Wiley and Sons, New York, pp. 125–128, Figure 128b.

14. Kani, G. N. J. (1969) "A Rational Theory for the Function of Web Reinforcement", *Journal of ACI* **66**, 186.
15. Park, R. and Paulay, T. (1975) *Reinforced Concrete Structure*, John Wiley and Sons, New York, pp. 280, 299, 307, 334.
16. Polaskus, M. N., Attiogbe E. K. and Darwin, D. (1981) "Shear Strength of Lightly Reinforced T-Beam", *ACI Journal* **78**, 450–451, Figure 3.
17. Kotsovos, M. D. (1983) "Mechanism of Shear's Failure", *Magazine of Concrete Research*, **35**(123).
18. Lampert, P. and Thurlimann, B. "Torsion Tests of Reinforced Concrete Beams (Torsionversuche an Stahbetonbalken)", *Bericht* No. 6506-2 Institut fur Baustatik, ETH, Zurich, June 1968, p. 101; and "Torsion Bending Tests on Reinforced Concrete Beams (Torsions-Beige-Versuche an Stahlbetonbalken)", *Bericht* No. 6506-3, Institut fur Baustatik, ETH, Zurich, January 1969, p. 99.
19. Hsu, T. T. C. (1982) "Is the 'Staggering Concept' of Shear Design Safe?", *ACI Journal*, No. 6 Proceedings **79**, 435.
20. Stamenkovic, H. (1979) "Short-Term Deflection of Beams", discussion, *ACI Journal*, No. 2 Proceedings **76**, 370–373.
21. Sears, F. W. and Zemansky, M. W. (1960) *College Physics*, 3rd edn, Addison-Wesley Publishing Company, London, England, Newton's Third Law, p. 20.
22. Timoshenko and Young (1940) *Engineering Mechanics*, McGraw-Hill Book Company, New York, pp. 5, 12.
23. McCormac, J. C. (1975) *Structural Analysis*, 3rd edn, Harper and Row, New York, p. 15.
24. McCormac, J. C. (1980) *Design of Reinforced Concrete*, 2nd edn, Harper and Row, New York, p. 191.
25. Rogovsky, D. M. and MacGregor J. G. "Design of Reinforced Concrete Deep Beams", *Developments in Design for Shear and Torsion*, Joint Committee ACI-ASCE 445, Phoenix, AZ, March 8, 1984.
26. Winter, G., Urquhart, L. G., O'Rourke, C. E. and Nilson, A. H. (1964) *Design of Concrete Structures*, 7th edn, McGraw-Hill Book Company, New York, pp. 68, 74.
27. Chu-Kia Wang and Salmon, C. G. (1965) *Reinforced Concrete Design*, International Textbook Company, Scranton, Pennsylvania, p. 68.
28. Ferguson, P. M. (1967) *Reinforced Concrete Fundamentals*, 2nd edn, John Wiley and Sons, New York, p. 168.

Chapter 4

Mechanism of vibrating fatigue failure of a reinforced concrete beam or any other member (a quantitative point of view)

4.1 A brief overview of the problem

As a result of the discovery of a new law in engineering mechanics, which governs flexural bending phenomena, it was possible to explain the mechanism of vibrating fatigue failure. Because of the simultaneous creation of internal active and internal resisting forces in a flexural member, the combination of internal active vertical shear forces and corresponding internal active compression and tensile forces leads directly to diagonal cracking and failure of a vibrating member.

Further, this chapter explains that a vibrating member in a second or higher mode causes points of inflection (PIs) where the portions on the left and right sides of each PI have a tendency to move in opposite directions. At the same time, it becomes clear that any vibrating member (for a given cycle of vibration) could be construed to be a continuous beam, uniformly loaded, and its safety could be calculated similar to that of a continuous beam. Yet, it is our intention to provide some qualitative, conceptual mechanism to explain vibrating fatigue. We do not intend to provide quantitative mathematical equations for fatigue.

It is clarified that PIs represent static supports for a series of simply supported beams, between adjacent PIs. Data are introduced for the first time concerning the role (function) of the PI in a continuous beam (or vibrating member), leading to an explanation of the mechanism of diagonal failure. It is proved that the stress condition of a continuous beam is fundamentally different from that suggested by Ritter and Morsch and a completely new concept is proposed for the structural calculation and safety of a continuous beam, with respect to diagonal failure.

This study suggests a new shear diagram which represents the real stress conditions of a continuous beam, with a clear distinction between the active shear forces and their corresponding resisting shear forces. Yet, with such differentiation of shear forces, the fallacy of the classical concept of diagonal tension is also proved.

4.2 Explanation of terms used in this study

If the member of a defined length and fixed end has shown only one curvature during vibration, one PI, called the second mode, will appear. If, in the course of vibration of the same member, two curvatures appear (the convex and the concave one), two PIs (called the third mode) will then appear; if three PIs, the fourth mode, and so on.

It is evident that the augmentation of the number of PIs, for a given length of the member, is conditioned by the augmentation of the force of vibration, decreasing the endurance of the vibrating member and leading to the failure, due to the fatigue of the material.

4.3 Introduction

The phenomenon of fatigue failure was noted by engineers as early as 1829, but the term "fatigue" first appeared in literature in 1854 in a journal of the Institution of Civil Engineers of England, in a paper by F. Braitwaite. The term is used for a variety of failures: mechanical, thermal, acoustic, or corrosion fatigue. This chapter is devoted to mechanical fatigue caused by repeated stressing due to mechanical forces which cause a member to vibrate.

The biggest percentage of mechanical failures of structural and machine parts, occurring during operation, can be attributed to fatigue. Many airplane accidents occur, apparently, due to vibrating fatigue failure of wing spars. Early jet airliner accidents were also attributed to fatigue failures. Structural fatigue has emerged as a major design problem for aircraft exceeding 30,000 flying hours.[1]

The failure of the Tacoma Narrows Suspension Bridge in 1940, after only four months in service, was caused by fatigue resulting from vibration due to wind action.

The lack of understanding of vibrating fatigue failure can best be seen from the existing definitions of this phenomenon: "Fatigue is the progressive change of structural and mechanical properties of metals produced by frequently repeated or fluctuating loading."[1] ... "Fatigue is a process of progressive permanent internal structural change in a material subjected to repetitive stresses."[2] In these definitions, however, no explanation is given of what forces are causing "progressive deterioration of the cohesion of metals." This study offers exactly such an explanation.

To understand the phenomenon of fatigue failure caused by vibration, one must first understand the mechanism of continuous beam failure through PIs[3,4,5] because, in essence, any vibrating element represents a continuous beam which oscillates in two opposite directions. The location of PIs is a function of exterior loads or vibrating forces. Numerous leading textbooks have been quoted, stating that any continuous beam is composed of a series of simply supported beams and emphasizing the fact that such static supports have been used to develop essential formulas for calculation of stresses in a continuous beam. (The fixed-end beam is the best example.)

4.4 Theoretical considerations

In the introduction in Chapter 2, it was stated that, in flexural bending, both Newton's third law and our new law are applicable simultaneously. Newton's third law is manifested by the fact that any action causes its own reaction. Physical support R_1 caused its own reaction R_{1r} (as illustrated in Figure 2.6); external load F causes its own reaction F_r; R_2 causes its own reaction R_{2r}. At the same time, the bending phenomenon causes additional internal active (compression, tensile, horizontal shear and vertical shear) forces and corresponding internal resisting (compression, tensile, horizontal shear and vertical shear) forces, which will be the main topic of discussion in this chapter (see Figure 2.6).

Internal forces in flexural bending have not been discussed by Newton or anyone before (Copernicus, Galileo) or after him. So, when considering these forces, no one has succeeded in suggesting the simultaneous existence of internal active and internal resisting forces up to now. Their physical existence has been recognized in textbooks dealing with strength of material (Timoshenko,[6] Saliger,[7] Bassin–Brodsky–Wolkoff,[8] Seely–Smith,[9] Kommers,[10] Winter–Nilson[11]) but no one has been able to rationally apply such forces for the equilibrium of a unit element (cube) cut out from a flexural member. This alone explains why the theory of diagonal tension[12] has not been understood. In addition, no one was able to rationally apply the same forces for equilibrium of a free body (as a portion of a flexural member). Consequently, it has been physically and mathematically impossible (until now) to prove that $\sum X = 0, \sum Y = 0, \sum M = 0$. If this had been understood, then the concepts of diagonal tension, the truss analogy theory, the same shear diagram for a fixed-end beam and a simply supported beam, and the concept of the algebraically unprovable equilibrium of a free body, would not have survived to this day!

4.5 Discussion

4.5.1 Diagonal cracking in a flexural member is caused by punch shear forces

Figure 4.1 shows that under the action of the external force R_1, the portion of the beam above the support has a natural tendency to move upward;[6,9,10] and the portion of the beam under the concentrated load has a natural tendency to slide downward as a result of the action of the external force P.[6] With the tendency of movement of the left and right portions in opposite directions, punch shear forces are created in the beam, leading to diagonal cracking, as shown in Figure 4.1(a).

Obviously, the resultant punch shear force V_n will be created for any type of load, concentrated, uniform, etc. As long as pure bending does not exist (where transversal forces V_1 and V_2 are eliminated), the resultant punch shear force V_n will always be present as a result of the combination of the vertical shear forces (V_1 and V_2) and flexural tensile forces (T). This will lead to the diagonal cracking of a bent member of a simply supported or continuous beam.[13]

Simultaneously with the action of punch shear forces V_n, sliding punch shear forces V_n' exist, acting parallel to diagonal cracking, one upward and another downward. These forces are caused by a combination of vertical shear forces V_1' and V_2' with compression force C, as shown in Figures 3.2 and 4.7(b). Evidently,

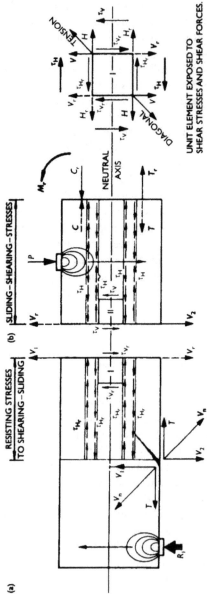

(a)

RESISTING STRESSES
|TO SHEARING—SLIDING|

SLIDING—SHEARING—STRESSES

(b)

NEUTRAL
AXIS

UNIT ELEMENT EXPOSED TO
SHEAR STRESSES AND SHEAR FORCES.

PUNCH SHEAR CRACKING IS CAUSED BY:

V_i = VERTICAL SHEAR FORCE CAUSED BY THE SUPPORT R_i; V_s = VERTICAL SHEAR FORCE CAUSED BY THE EXTERNAL LOAD P;
T_i = FLEXURAL TENSILE FORCES CAUSED BY BENDING; V_n = RESULTANT PUNCH SHEAR FORCES CAUSED BY THE COMBINATION
OF SHEAR FORCES V_1 AND V_2 WITH FLEXURAL TENSILE FORCES T.

NOTES

τ_V = VERTICAL RESISTING SHEAR STRESSES (AGAINST SLIDING) CAUSED BY THE STRENGTH OF THE MATERIAL. τ_H = HORIZONTAL RESISTING
SHEAR STRESSES (AGAINST SLIDING) CAUSED BY COHESION OF THE MATERIAL. τ_{V_r} = VERTICAL SLIDING SHEAR STRESSES CAUSED BY SLIDING
SHEAR FORCES. τ_{H_r} = HORIZONTAL SLIDING SHEAR STRESSES CAUSED BY SLIDING SHEAR FORCES. C_r = RESISTING COMPRESSIVE FORCES
CAUSED BY RESISTANCE OF THE MATERIAL TO COMPRESSABILITY. T_r = RESISTING TENSILE FORCES CAUSED BY THE RESISTANCE OF THE
MATERIAL TO STRETCHING.   C = INTERNAL COMPRESSIVE FORCES CAUSED BY BENDING FORCE P.   T = INTERNAL TENSILE FORCES CAUSEDBY
BENDING FORCE P. M_r = RESISTING MOMENT (CAUSED BY THE RESISTING FORCES) TO BALANCE THE EXTERNAL MOMENT AT THE GIVEN
CROSS SECTION.

Figure 4.1 This figure illustrates (1) horizontal resisting shear stresses τ_{H_r} (which prevents sliding) as shown by Timoshenko,[6] Winter and Nilson[11] and horizontal sliding shear stress τ_H (which causes sliding), as shown by Saliger[7]; (2) vertical resisting shear stresses τ_{V_r} (which prevents sliding) and vertical sliding shear stresses τ_V (which causes sliding), as shown by Saliger[7]; (3) diagonal cracking caused by resultant punch shear forces V_n; and (4) that probable diagonal tension of a unit element is in the opposite direction to the Ritter–Morsch diagonal.

diagonal cracking is caused by punch shear forces V_n and not by any diagonal tension, as is seen clearly in Figures 3.2, 4.1(a) and 4.7. By proving that diagonal cracking in a flexurally bent beam is caused by punch shear forces V_n, the fallacy of the Ritter–Morsch concept of diagonal tension is also proved, which Figures 4.1(a) and (b) indisputably demonstrate. See also Figure 4.2.

4.5.2 Stress condition at the inflection point of a vibrating member

In order to understand the phenomenon of fatigue failure caused by vibration, let us investigate a deformed leaf spring (which simulates the condition of a flexed member at the peak deflection of a vibration cycle) or the deformation of a very elastic continuous beam.

Assume a very thin leaf spring anchored at its ends (anchorage points 1 and 4) and deformed by forces F_1, F_2, F_3, as shown in Figure 4.3. After such deformation, forces F_1, F_2 and F_3 are fixed to remain at the positions shown. Deformation is similar to semi-circles.

There is no doubt that such a deformed portion of a spring, located in field 1, has a natural tendency to separate from the portion located in field 2, precisely at the PI denoted by 2. Such a tendency for separation and division also exists at the PI 3 between fields 2 and 3. It is obvious that bending force F_1 tends to cause the spring to shear through PI 2 and move the right portion in field one upward. Such separation is special vertical shear, caused by two (static) supports of two oppositely bent simply supported beams, which has no correlation with the so-called diagonal tension. At any other cross section, transversal shear exists caused by the support (V_1) and the external load (V_2) as shown in Figure 4.1(a) and references 3–5, 8, 14–16.

4.5.2.1 Vertical shear forces at the PI and at any other cross section

The differences between vertical shear forces at any cross section and the vertical shear forces at a PI are so profound that they are almost not comparable:

1   At any cross section, we deal with vertical shear forces of a simply supported beam, bent in one direction, where a shear force (V_1) is caused by one support and oriented as the support itself, while another force (V_2) is caused by the external load and oriented as the load itself. The combination of these two forces with flexural tension produces punch shear forces (V_n), which cause diagonal cracking as shown in Figure 4.1(a).

2   On the other hand, in Figure 4.8 at the cross section at the PI, we are dealing with two simply supported beams, oppositely bent, usually with different spans and different stressed conditions; connected at the inflection plane by cohesion of the material, but with two oppositely oriented vertical shear forces

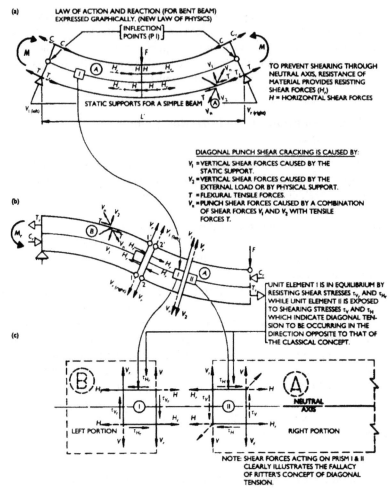

(a) INSTEAD OF FORCE F, SAME BENDING CAN BE REACHED BY ECCENTRICAL COMPRESSIVE FORCES (C) & TENSILE FORCES (T). RESISTANCE OF MATERIAL TO DEFORMATION INTRODUCES RESISTING COMPRESSIVE FORCES (C_r) & RESISTING TENSILE FORCES (T_r). SHEAR FORCES ARE NOT PRESENT FOR PURE BENDING.

(b) V_r = VERTICAL RESISTING SHEAR FORCES; H_r = RESULTANT OF RESISTING HORIZONTAL SHEAR FORCES; T_r = RESULTANT OF TENSILE RESISTING FORCES; C_r = RESULTANT OF COMPRESSIVE RESISTING FORCES; V_1, V_2 = VERTICAL SHEAR FORCES; H = HORIZONTAL SHEAR FORCES; τ_{V_r}, τ_{H_r} = RESISTING SHEAR STRESSES. (BY REF.6 AND REF.12); τ_V, τ_H = ACTIVE SHEAR STRESSES. (BY SALIGER-REF.7).

(c) FOR EQUILIBRIUM OF ANY FREE BODY (BEAM Ⓐ OR INFINITESIMALLY SMALL PRISM ①) LAW OF ACTION AND REACTION MUST BE USED.

Figure 4.2 Distribution of resisting forces in the immediate vicinity of the inflection plane for equilibrium of any portion of a bent beam, only law of resisting forces is applicable. Portion (B) is not held in equilibrium by the forces acting on portion (A), rather by forces replacing the strength of cut-off fibers of portion (B).

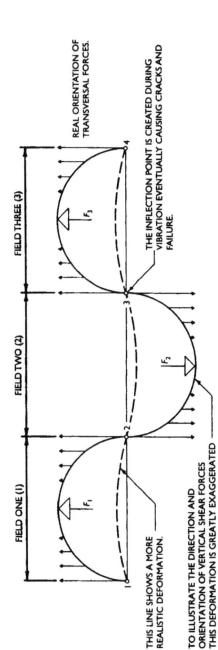

Figure 4.3 This figure represents a spring with fixed ends (points 1 and 4) deformed by forces F_1, F_2 and F_3. At points 2 and 3, the natural tendency to separate field 2 from fields 1 and 3 exists. Such a division or shearing exists only at the PIs. The tendency to shear is responsible for the fatigue failure caused by vibration.[20]

$(V_l$ – left and V_r – right) which are caused exclusively by the two supports and directed (oriented) as static supports themselves.

3   In the immediate vicinity of the inflection plane, to the left- and right-hand side, punch shear forces V_n act, created by vertical shear forces V_1 (caused by the support) and vertical shear forces V_2 (caused by the external loads of a given simple beam) in combination with flexural tensile forces T (as illustrated by Figures 4.7(a) and (b)). This leads to diagonal cracking and eventual fracture. Yet, vertical shear forces at the PI never combine with any other internal or external forces.

4   Vibrating forces could be shown as a uniform load, so vertical shear forces V_1 and V_2 (for a given simple beam) are the largest in the immediate vicinity of the (static) support. Consequently, critical punch shear forces V_n are always present in the vicinity of the PI of a vibrating member: the larger the punch shear forces V_n, the shorter the life of a vibrating member.

5   Additional differences between shear forces at any cross section (V_1 and V_2) and at the PI (V_l – left and V_r – right) lie in their reactions (V_r) where, *at any cross section*, resisting shear forces V_r prevent shearing-sliding at that cross section and also serve as forces for equilibrium of a portion of a free body, while the resisting shear forces (V_r) *at the PI* serve exclusively to prevent sliding–shearing of the two beams in opposite directions at the PI.

6   Furthermore, PIs in a continuous beam can be considered as theoretical supports for simple beams located between two adjacent PIs.[8,14] The reactions of the two adjacent simple beams are located in one and the same cross section, which passes through the PI, but in two planes infinitely close to each other. These two reactions (resisting shear forces V_r as strength of materials) are of equal intensity, but opposite in orientation, and are caused by the action of vertical shear forces at the inflection plane.[3–5,8,14,16–18]

4.5.2.2   Vibration fatigue failure and internal forces

In Figure 4.4, in field 1, all transversal forces are caused by the external load and are oriented upward following the orientation of bending force F_1 and the bending line itself, while all vertical shear forces caused by the supports are oriented in a direction opposite to the bending forces.[13]

It is the same with field 2; all transversal forces follow the orientation of bending force F_2 while the reaction (supports) are oriented oppositely. Reversal of transverse forces caused by the external loads exist only at the PIs 2 and 3. At any other cross section, transversal forces (caused by the external loads) are oriented up or down depending on the direction of the respective load. The fact that the direction of transverse or vertical shear forces (caused by the external load) follows the orientation of the bending line of any loaded or vibrating member, should be particularly emphasized. Because oppositely bent portions of any continuous beam are connected at the inflection planes and are in equilibrium, it is concluded that,

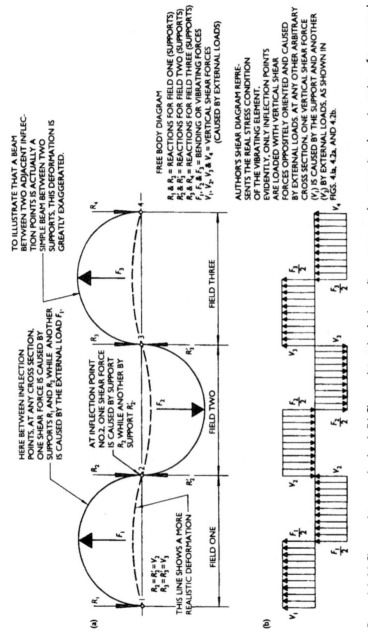

Figure 4.4 (a) Shows a beam under load. (b) Shows the author's shear diagram correctly portraying the orientation of transversal forces. Opposite vertical shear forces, which exist only at the PIs, are responsible for any fatigue failure created by vibration. The shear forces shown are caused exclusively by external loads. Vertical shear forces caused by supports will be oriented as supports — here oppositely to vertical shear forces shown.

at the PIs, two oppositely oriented vertical shear forces must exist, as is recognized in literature.[8,14,19–22]

As a result of the existence of such forces in the contraflexural area of a concrete member, failure due to vertical shear forces V_1 (left) and V_r (right) at the PI is possible by shearing (sliding through the inflection plane) in the opposite direction of two simply supported beams.[23] The presence of longitudinal reinforcement will impose punch shear failure,[24] (as happened with the rigid concrete frame in Shelby, Ohio, in 1955).[3,25,26] This is similar to the illustrations in Figures 4.1, 4.2 and 4.7. This means that statical support (or vertical shear force V_1 – left) develops punch shear cracking in its flexural tensile zone, as any physical support or pullout test shows (see references 27–29 for a more complete depiction of this phenomenon). Such cracking is at approximately 45° in slope. At the same time, the right statical support (V_r – right) develops punch shear cracking (in its own tensile zone) at approximately 45° in slope, as illustrated in Figure 4.8.

We do not question the signs and symbols of the classical shear diagram. What is in question is the real direction of shear forces and the real stress condition of a deformed continuous beam, which was not recognized by the shear diagram of the classical theory of elasticity. This peculiar stress condition at the PI, with oppositely oriented reactions of two simply supported beams, is undoubtedly the most sensitive and most critical cross section of a deformed beam or vibrating member, because of the natural tendency of such members to separate by division through a PI and move in opposite directions: "In thin web – shear cracking occurs at PI of continuous beam."[30] Support for this statement is apparent from Figures 4.3 and 4.4 and references 5, 8, 14 and 19–21 – Figure 5. Our new shear diagram in Figure 4.4(b) is supported by an identical shear diagram as shown by Figure 5 in reference 21, Figure 2c of ACI Committee 442,[19] and (above all) by common sense!

Finally, let us look at the same beam shown from the point of view of the classical concept, but with the beam illustrated as shown in Figure 4.5.

It immediately becomes clear that Figure 4.5 actually represents the same spring as shown in Figure 4.3. The cantilever loads at the ends (1 and 4) of the beam in Figure 4.5 are the equivalents of the anchorage points 1 and 4 in Figure 4.3, and forces (F_1 and F_3) in Figure 4.5 are the equivalents of bending forces F_1 and F_3 in Figure 4.3. Most importantly, these forces (F_1 and F_3) are, in reality, reactions or physical supports R_1 and R_2, respectively. Also, the fact should be emphasized that any physical supports in the field of negative moments are bending forces in a continuous beam, while PIs are static supports. "AISC and AASHO specifications suggest that center portions of continuous spans may be treated as simple spans between points of contraflexure, while portions over the supports may be treated similar to cantilevers (Figure 9.19)."[31] The quoted figure called the portions of the beam between two PIs "equivalent simple spans," and the portions of the beam above the supports, "equivalent cantilever spans," where the free end of a cantilever is defined as "a point of contraflexure," (reference 32, p. 242).

The above statement of AISC and AASHO that any continuous beam is composed of a series of simply supported beams could be easily realized from the following illustration.

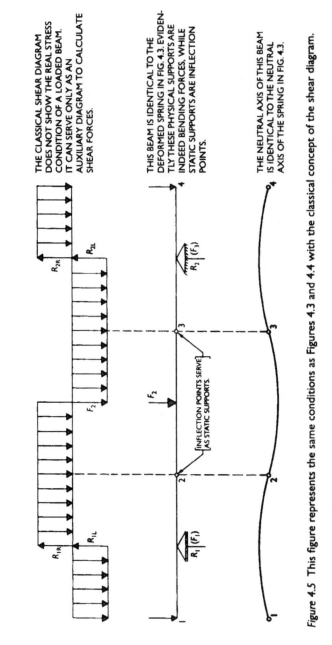

THE CLASSICAL SHEAR DIAGRAM DOES NOT SHOW THE REAL STRESS CONDITION OF A LOADED BEAM. IT CAN SERVE ONLY AS AN AUXILIARY DIAGRAM TO CALCULATE SHEAR FORCES.

THIS BEAM IS IDENTICAL TO THE DEFORMED SPRING IN FIG. 4.3. EVIDENTLY THESE PHYSICAL SUPPORTS ARE INDEED BENDING FORCES, WHILE STATIC SUPPORTS ARE INFLECTION POINTS.

THE NEUTRAL AXIS OF THIS BEAM IS IDENTICAL TO THE NEUTRAL AXIS OF THE SPRING IN FIG. 4.3.

Figure 4.5 This figure represents the same conditions as Figures 4.3 and 4.4 with the classical concept of the shear diagram.

At any support of a simply supported beam, a shear force exists, caused by the support and oriented as the support itself. Assume that the direction is vertical. If one rotates such a flexurally bent beam vertically by 180°, the vertical shear forces, caused by the support, must be oriented downwardly. Rotation of a flexurally bent member does not and cannot influence the correlation between the support and the direction of action of its shear forces: shear forces caused by the support must be oriented as the support itself, downwardly. Yet, if we put together two identical, but oppositely bent, simply supported beams close to each other in such a way that they touch each other at their ends (not glued together), the shear forces of these two beams must be oppositely oriented as their supports. If we now eliminate the two closest columns as supports and fix them at that touching vertical plane by special glue (say epoxy), they will remain as two simply supported beams with the same oppositely oriented vertical shear forces, changing only the name: instead of two simply supported beams, we call this configuration a continuous beam.

Evidently, vertical shear forces, which change the direction of action now at an inflection plane (which replaced two physical supports), are a simple law of engineering mechanics. This is graphically illustrated in Figure 2.11, where shear forces at the inflection point (plane) must be oppositely oriented. This fact of the existence and action of two vertical shear forces, oppositely oriented at static support (PI), is also well illustrated by references 3–5, 8 and 16.

It is clear that the classical shear diagram fails to represent the real stressed condition of a continuous beam, while our shear diagram (Figure 4.4) does.[3-5] This is the basis for comprehending the concept of vibrating fatigue failure. By a brief inspection of our shear diagram, the importance and role of PIs in causing fatigue failure by vibration can be seen clearly. With such recognition of the actual stress conditions at PIs, it is not difficult to understand the contribution of this essential factor, which leads to fatigue failure in a vibrating member, as is illustrated in Figure 4.7.

From another point of view, in a structural member which can vibrate in a second or higher mode, a PI must be treated as a hinge for hanging two oppositely bent members[4] ("where one part would be moving in one direction, while another part would be moving in the opposite direction").[20] These members should not separate by division through the PI. Yet, the entire portion of the vibrating member, where a PI could occur, must be treated as a hinge capable of holding two parts together without their separation at the PI.[4,5] This is applicable to airplanes, boats, submarines, nuclear structures, bridges, buildings or its members, which can vibrate in a second or higher mode.

4.5.3 Simple experiments concerning vibrating fatigue failure

The phenomenon of vibrating fatigue failure can be demonstrated in a simple experiment with a piece of copper wire, as illustrated in Figure 4.6. The copper wire has a diameter of 2 mm and its ends are fixed. Loading is imposed by the

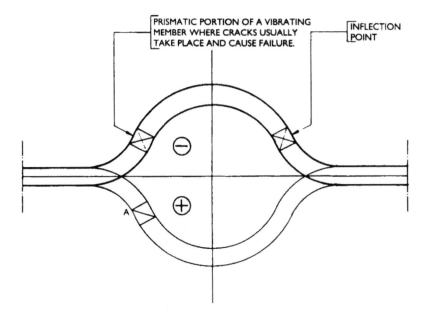

Figure 4.6 A typical example of fatigue failure: vibration produces the same deformation as static loads imposed alternately up and down.

fingers, for bending and deflection of the wire in alternate directions. This bending is very similar in form to a vibrating member with a constant load.

If the wire is alternately deflected or oscillated at the rate of one cycle per second, after approximately 20 s, the temperature developed in the regions of the PIs can be tolerated by the fingers; but around the 30th second, the temperature developed becomes so high that the fingers cannot tolerate further bending and the experiment must be interrupted for a few seconds so that the wire can cool. With about 40 cycles of bending, the wire failed through one of the PIs.

In a further test with a wire of the same diameter, same length, same number of oscillations per unit time, but with a smaller load imposed (resulting in shallower deflections) about 150 oscillations took place before fatigue failure, or rupture occurred. The developed temperature was tolerable for the fingers throughout the experiment so that bending was continuous until the wire suddenly ruptured.

Again, using a third wire with similar characteristics as the two previous wires and one bending cycle per second, but with a much smaller load (expressed in much larger radius of bending curvature) about 500 oscillations were reached before rupture took place. The temperature rise was almost undetectable.

The stress condition of a bent member (wire) is a function of its deflection curve where any change of deflection (bending), from concave to convex, causes changes in the orientation of vertical shear forces in the beam. By alternating the bending

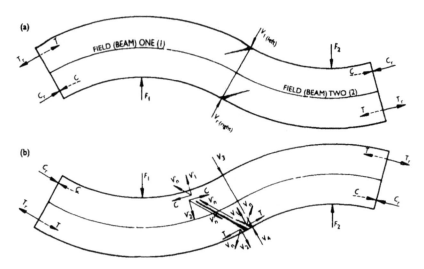

Figure 4.7 Mechanism of vibrating fatigue failure due to PIs as a result of (static) supports action at the PI in a close vicinity to the inflection plane (at the left and right sides) in flexural tensile zones; there exists a permanent action of punch shear forces V_n produced by a combination of vertical shear forces V_1 (caused by the static support) and V_2 (caused by the external load) with flexural tension forces T, as illustrated in this figure and Figure 4.1(a).

direction and increasing the load, punch shear forces V_n (Figure 4.7) produce cleavage stresses which alone can cause plastic deformation in the critical prism (the portion containing the PI, as shown in Figure 4.6), while all other stresses are still in the elastic stage. As a result of the plastic deformation, diagonal cracking is caused. Finally, cracks occur and progress along the diagonals of the critical prism.

We state again that at any cross section, two active vertical shear forces are developed simultaneously; one caused by the support and oriented as the support and another caused by the external load and oriented as the load itself. Both vertical shear forces are combined with tensile forces T, leading to punch shear cracking (as illustrated in Figures 3.2 and 4.7(b)).

In Figure 4.7(a), vertical shear force V_1 (left) is the static support for the left portion, causing its own punch shear cracking. In Figure 4.7(b), V_4 is the static support for the left portion causing its own punch shear cracking. So, during vibration, punch shear cracking, caused by static support V_4, unites its crack with the cracking caused by punch shear force V_n'. This is caused by a combination of active internal vertical shear forces V_i' with active compression forces C, producing oppositely oriented sliding forces V_n' (parallel to cracking caused by V_l (left) and V_r (right), which accelerates a diagonal failure). Further, by conversion from a concave to a convex position during vibration, the entire prism covered by 45° diagonals becomes exposed to cracking from both sides.

As the sound cross section of the member is reduced to the point that it becomes incapable of withstanding the action of the vertical shear forces as static support, the member suddenly fails in punch shear.[33,34]

From what has been stated, it follows that diagonal failure through the critical prism is the natural tendency of a deformed beam to release itself from its deformed shape and stressed condition, by division. Therefore, to understand fatigue failure caused by vibration more clearly, we introduce Figure 4.8, which shows the stress condition in the vicinity of the PI (critical prism) of a loaded member.

As can be seen, horizontal shear forces, as well as horizontal compression and tensile forces, are zero at the inflection plane. The only forces acting at that cross section are vertical shear forces, represented by shear forces V_l (left) and V_r (right), which are reactions (statical supports) to the vertical loads of left and right simply supported beams. The forces V_l and V_r have a tendency to cause rupture while the internal resisting vertical shear forces V_r (as strength of materials) have a tendency to prevent such rupture, precisely through the inflection plane, by preventing the sliding of two beams in opposite directions[4] (see Section 4.5.2.2 for more detail).

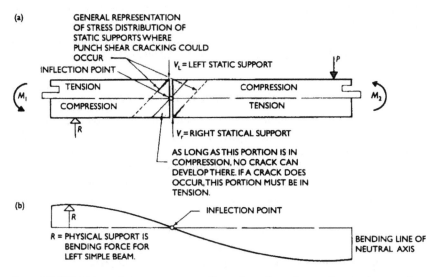

Figure 4.8 This figure represents a contraflexural portion of a continuous beam with a specific area where diagonal cracking could be developed by V_l (left) and V_r (right) as shown in (a).

So, for any cracking or rupture through the inflection plane, only vertical shear forces can be responsible, because there are no other forces at that cross section. "If vertical reinforcement of shear wall is weak, a sliding shear failure . . . may occur."[35]

Evidently, initial cracking in a metallic structural member may originate in an early stage of vibration but cannot be detected by the naked eye. However, it can be detected using X-ray procedures.[36] Real failure takes place some time after the first crack comes into existence. Such cracks can exist for months or years before fracture takes place. Because cracks occur long before fracture takes place,[37] it is not brittle failure, but rather progressive, undetected cracking leading to failure.[13,38]

Many textbooks state that repeated stressing influences the crystals of materials, causing internal structural changes. But this is not true where vibrating fatigue failure is concerned. There is no interrelation between the number of oscillations and the structural changes in the crystals. Structural changes in the crystals are not influenced by vibratory movements of the member and cannot be influenced unless the temperature of the member reaches a critical value: "Crack appears to be non-crystalline in nature and is a function of the macroscopic plastic strains."[36]

The statement of previous researchers that the surfaces of materials are much weaker in fatigue than the underlying material[39] is also an erroneous statement. Actually, the opposite is true. The surface of a metal is much stronger than the underlying material. This can be deduced by comparing thin and thick steel wire exposed to fatigue failure, where it was found that thinner elements have higher resistance to fatigue failure than thicker ones.[2] This is because the ratio of peripheral surface area to cross-sectional area is higher for thin elements in comparison to thicker elements. Researchers arrived at the erroneous conclusion because cracks always occur first at the peripheral surface of the member and because the leading theory of fatigue failure was based upon "structural changes of material."

4.5.4   The main contributors to fatigue failure

In addition to the aforementioned essential contributing factor, other contributing factors exist. One of these is the imperfect elasticity of structural materials. After unloading, a given member never regains its prior shape and dimensions.[36,39,40] This is caused by residual microscopic stresses produced by loading of such members. Many of these microscopic residual stresses are of the same orientation as the main stresses so they will add to the main stresses in the region of the residual stresses.

Such addition or accumulation of stresses, by repeated loading and unloading (as is the case with vibrating members), can cause some of the exterior fibers to reach a plastic stage. This, in turn, leads to the creation of fissures and cracks. The existence of even the smallest amount of residual stress in some region will cause that unloaded member to retain residual deformation. It becomes permanently shorter or longer depending upon loading, compression or tension. In a beam, where

PIs do not exist, the accumulation of residual stresses occurs at the region where they originate. In the member with PIs, which serve as anchors for corresponding beams, the residual stresses concentrate in the region of the PIs because the distance between adjacent PIs is constant, and because the cohesive forces of the material have a tendency to resist change in the distance between adjacent PIs. The reason for stress concentration at the PI is that when, during a cycle, the prior compressed fibers pass into tension, the portion between adjacent PIs can never return to its prior length. Because these fibers of the vibrating member (now in tension) are connected with the compressed fibers precisely at the inflection planes, the member is not free to remain shorter and due to existing cohesive forces (or strength of materials) at the inflection planes, the portion between the vertical planes at the PI will be stressed more than the rest of the member. Therefore, stresses accumulate and concentrate in the region of the inflection plane as a result of reversal loading, sudden stretching and sudden compression of this portion. Sudden stretching or compressing, such as a dynamic load, stresses the area nearest its impact the most. The nearest area is in the vicinity of the inflection planes which serve as anchors for compressed and tensioned fibers. These compressed fibers have the natural tendency to prevent stretched fibers from becoming shorter. With continued vibration, residual micro-stresses build up until they reach the plastic stage in the exterior fibers of the vibrating member. Plastic deformation of any fiber will soon lead to fissures and cracking of the member in the vicinity of the inflection plane. Stress concentration immediately occurs at any crack. Also, there are inherent flaws and defects in any material, which might occur during the manufacturing process, or which the material will develop at some stage of its life.[36] Stress concentration deepens the cracks, thereby decreasing the sound cross section, until the section becomes incapable of withstanding the imposed load and fracture takes place due to vertical shear or eventually due to punch shear.[37]

Corrosion fatigue originates from stress conditions which concentrate on the surface of a metal around any uneven spot (such as a pit, small hole or cavern) caused by corrosion. Such a stress concentration leads to a plastic stage of the material around the uneven spot, initiating fissures and cracking.[37] On the basis of what has been said, it follows that the most critical area of such corrosion for an operating member is in the vicinity of the inflection plane because of the inherent nature of the contraflexural area for stress concentration caused by other factors.

The plastic state (leading to a fracture) can be created by a combination of the applied stresses, local stress concentrations, residual stresses and dynamic stresses; any or all of them, acting simultaneously.

In fatigue failure, three different factors act concurrently: cleavage forces (or punch shear forces) which decrease the sound cross section of the member; plastic deformation, which penetrates deeper into the cross section at the PI; and temperature increase, which decreases yield strength. Thus, it is clear that the PI is a nucleus about which a group of factors act simultaneously during vibration of the member and lead to fatigue failure. Therefore, by eliminating PIs, which can be achieved in several ways, vibrating fatigue failure can be avoided.[15]

By different means and methods, a given member could be forced to vibrate as we desire; first, second, third or higher mode. Evidently when we change the mode of vibration we change the duration of the vibration. This could be achieved by prestressing, poststressing, by the combination of two or more different materials, by replacing one material with another, by changing the dimensions or the cross section, and so on.

4.6   Conclusion

From the foregoing analysis and discussion, the following conclusions can be drawn:

1   While in a pure push–pull concept, internal forces could only be resisting forces in a flexural member, the internal resisting forces of the push–pull concept coexist together with internal active and internal resisting forces created by flexural bending.

2   In order to establish any equilibrium of a free body, the forces themselves must be in equilibrium and, as a result of their equilibrium, the equilibrium of a free body will be possible: $V = V_r, C = C_r, T = T_r, H = H_r, R_1 = R_{1r}, F = F_r$ (Figures 2.6 and 5.1).

3   The existing concept for equilibrium of a free body (for example $\sum X = 0$) does not stipulate the real equilibrium of horizontal forces, because they are not located on the same line of action, even though that so-called compression force C is equal to tensile force T. Consequently, they can never be in equilibrium.

4   Also, the equilibrium of any free body would be possible if it is algebraically provable that $\sum X = 0, \sum Y = 0$ and $\sum M = 0$, which was unprovable until now.[14] Our concept fills such a vacuum in structural engineering theory, so it will not be "merely statements of Sir Isaac Newton,"[14] but rather the law of mechanics.

5   Vertical shear forces used for equilibrium of a portion of the free body, as shown in most textbooks, are not shear forces V but rather resisting shear forces V_r, as shown by Timoshenko,[6] Seely–Smith[9] and Kommers.[10]

6   The same vertical resisting shear forces are applied for equilibrium of a portion of the free body diagram as those for a unit element, as shown in Figures 4.1(a) and (b). Such forces can prevent shearing of the beam at any cross section, thus they cause equilibrium of a portion of the free body diagram (or unit element), but they never cause any deformation of a beam, portion of a beam, or unit element (stretching, compression or any diagonal tension).

7   By applying *active* shear stresses on a unit element (as shown by Timoshenko,[6] Seely–Smith,[9] Saliger,[7] Winter–Nilson[11] and by Figure 4.1(a)), diagonal tension is possible but in the opposite diagonal to the Ritter–Morsch diagonal, as is illustrated in Figures 4.1(b) and 4.2(c) – Prism II.

8 Evidently, in a continuous beam, two fundamentally different vertical shear forces exist simultaneously:

   a   vertical shear forces at any cross section, one which is caused by the supports (and oriented as the supports) and the other which is caused by the external load (and oriented as the load itself); and
   b   two oppositely oriented vertical shear forces at the PI, where both these forces are caused by the supports (or by two simply supported beams), and are oriented opposite to the bending forces, so that the PI is only a cross section where bending forces change direction of action (Figure 4.8(a)).

9 The curvature of the neutral axis of a deformed beam is the only natural guide to the orientation (direction) and magnitude of shear forces caused by the external load, their resisting shear forces, and the shear diagram.

10 Any diagonal cracking of a flexural bent member is caused by pure punch shear forces V_n, produced by a combination of vertical shear forces V_1 (caused by the support), vertical shear forces V_2 (caused by the external load), and flexural tensile forces T, as illustrated in Figure 4.1(a).

11 Because vibrating forces could be shown as the action of uniform loads, it follows that the resultant punch shear forces V_n are the largest in the immediate vicinity of the supports (or at the PI as static supports), where the first punch shear cracking is expected to develop. This is proven by diagonal cracking in common practice.

12 It becomes evident that the mechanism of vibrating fatigue failure of a member is the result of punch shear cracking, caused by the static supports (V_l left and V_r right), in combination with cracking caused by punch shear forces V'_n, as illustrated in Figures 3.2, 4.7(a) and (b).

13 The vibrating fatigue failure of the metallic material is initially progressive, undetected cracking and later, sudden fracture due to punch shear but not brittle fracture.[40]

14 The non-ideal elasticity of any material is probably the greatest contributor to fatigue failure because in the reverse cycle of the vibration, the compressed zone cannot return to its original length when the zone becomes tensioned, and microscopic residual stresses will develop in the vicinity of the inflection plane.

15 Fatigue failure of a vibrating member is not caused by tension (as generally stated in textbooks) but by punch shear forces created by a combination of vertical shear forces with flexural tensile forces.

16 Vibrating fatigue failure is also not caused by any structural changes of the material, deterioration of cohesion, or by the weaker surface of metal, but rather, by punch shear failure, as shown in Figure 4.7. ("In fact, under the second and third mode, one part of the building would be moving in one direction while another part of the building would be moving in the opposite direction."[20] ... "horizontal slip in some level."[22])

References

1. *Encyclopedia Britannica* (1970) William Benton, Chicago, Book 9, p. 113.
2. American Concrete Institute Committee 215 (1974) "Consideration for Design of Concrete Structures Subjected to Fatigue Loading", *Journal of the ACI*, Proceedings **81**, 97.
3. Stamenkovic, H. (1977) "Inflection Points as Vertical Statical Supports are Responsible for Structural Failure of AMC Warehouse in Shelby, Ohio, 1955", *Materials and Structures* **10**(60), 375–384.
4. Stamenkovic, H. (1979) "Simplified Method of Stirrups Spacing", discussion, *ACI Journal* 68–71, Figures A and B.
5. Stamenkovic, H. (1979) "Specifying Tolerance Limits for Meridional Imperfections in Cooling Towers", discussion, *ACI Journal* 1023–1029.
6. Timoshenko, S., MacCullough and Gleason H. (1961) *Elements of Strength of Materials*, 3rd edn, 6th printing, D. Van Nostrand Company, New York, pp. 94, 95, 137, Figures 111, 159, 160.
7. Saliger, R. (1927) *Praktische Statik*, Franz Deuticke, Leipzig and Wein, p. 148, Figure 198.
8. Bassin–Brodsky–Wolkoff (1969) *Statics and Strength of Materials*, 2nd edn, McGraw-Hill Book Company, p. 250, Figures 11.4, 11.5, 11.6.
9. Seely, F. B. and Smith, J. O. (1956) *Resistance of Materials*, 4th edn, John Wiley and Sons, New York, pp. 125–128, Figure 128.
10. Kommers, J. B. (1959) "Mechanics of Materials", *Civil Engineering Handbook* editor-in-chief, Leonard Church Urquhart, New York, pp. 3–34, Figure 43.
11. Winter G. and Nilson, H. A. (1973) *Design of Concrete Structures*, McGraw-Hill Book Company, New York, pp. 62–63, Figures 2–13 and 2–14.
12. Chu-Kia Wang and Salmon, C. G. (1965) *Reinforced Concrete Design*, International Textbook Company, Scranton, Pennsylvania, p. 63.
13. Stamenkovic, H., "Quantitative Evaluation of Shear Strength in a Flexural Member", *Developments in Design for Shear and Torsion*, Symposium paper, Annual Convention, Phoenix, AZ, March 8, 1984.
14. McCormac, J. C. (1975) *Structural Analysis*, Harper and Row, New York, pp. 15, 22, 283, 286, Figure 16-9a–Figure 16-10.
15. Stamenkovic, H., "Means for Fatigue Proofing a Solid Rotary Shaft", United States Patent #3811295, Washington, May 21, 1974.
16. Stamenkovic, H. (1979) "Short Term Deflection of Beam", discussion, *ACI Journal* 370–373.
17. Stamenkovic, H. (1978) "Suggested Revision to Shear Provision for Building Code", discussion, *ACI Journal* **75**, 565–567, Figure Ca.
18. Stamenkovic, H. (1978) "Suggested Revision to ACI Building Code Clauses Dealing with Shear Friction and Shear in Deep Beams and Corbels", *ACI Journal* **75**, 222–224.
19. ACI Committee 442 (1971) "Response of Buildings to Lateral Forces", *ACI Journal* **68**, 87, Figure 2c.
20. Amerhein, J. E. (1978) *Reinforced Masonry Engineering Handbook*, 3rd edn, Masonry Institute of America, Los Angeles, CA, p. 56.
21. Croll, T. G. A. and Kemp, K. O. (1979) "Specifying Tolerance Limits for Meridional Imperfections In Cooling Towers", *ACI Journal* **76**, 155, Figure 5.

22. Peter Mueller, "Behavioral Characteristics of Precast Walls", ATC-8 Seminar, Applied Technology Council, Berkeley, CA, April 27–28, 1981, p. 283.

23. Fuller, G. R. "Earthquake Resistance of Prefabricated Concrete Buildings State of Practice in United States", ATC-8 Seminar, Applied Technology Council, Berkeley, CA, April 27–28, 1981, p. 134.

24. Pinkham, C. W. and Moran, D. F. (1973) *"San Fernando, California, Earthquake of February 9, 1971"*, U. S. Department of Commerce, Vol. I, Part A, p. 292.

25. Elstner, R. C. and Hognestead, E. (1957) "Laboratory Investigation of Rigid Frame Failure", Symposium on AMC Warehouse Failure, *Journal of the ACI* 54, 554, 664, 667.

26. McKaig, T. H. (1962) *Building Failures*, McGraw-Hill Book Company, New York, pp. 180–181.

27. Cummins, A. E. and Hart, L. (1959) "Soil Mechanics and Foundation", *Civil Engineering Handbook*, 4th edn, ed Leonard Church Urquhart, McGraw-Hill Book Company, New York, pp. 8–10, 8–11, Figures 5, 6.

28. ACI-ASCE Committee 326. (1967) "Shear and Diagonal Tension", Proceedings, *ACI*, Vol. 59, January–February–March, p. 426.

29. Gustave, F. (1979) *Shear and Bond in Reinforced Concrete*, Trans. Tech. Pub., Boalsburg, Pennsylvania, p. 15, Figure 23.

30. Winter, G., Urquhart, L. C. O'Rourke, C. E. and Nilson A. H. (1964) *Design of Concrete Structures*, 7th edn, McGraw-Hill Book Company, New York, pp. 66, 73.

31. Bresler, B. and Lin, T. Y. (1963) *Design of Steel Structures*, John Wiley and Sons, New York, p. 459, Figure 9-19; p. 491, Figure 7-51.

32. Park, R. and Paulay, T. (1975) *Reinforced Concrete Structures*, John Wiley and Sons, New York, p. 242.

33. Stamenkovic, H. (1980) "Design of Thick Pile Caps", discussion, *ACI Journal* 77, 476–481.

34. Forsyth, P. T. E. "The Physical Basis of Metal Fatigue", American Elsevier Publishing Company, Inc., 1969, reprinted in *Annotated Bibliography on Cumulative Fatigue Damage and Structural Reliability Models*, System Development Corporation, USA Department of Commerce, p. 106, paragraph 1.

35. Park, R. and Paulay, T. (1980) *Concrete Structures*, "Design of Earthquake Resistant Structures", ed E. Rosenblueth, John Wiley and Sons, New York, Chapter 5.

36. Frost, N. E., Marsh, J. J. and Pook, L. P. (1974) *Metal Fatigue*, Clarendon Press, Oxford, England, pp. 4, 455, 459, 463, 222.

37. Munse, H. H. (1968) "Fatigue and Brittle Fracture", *Structural Engineering Handbook*, ed Edwin H. Gaylor, Jr. and Charles N. Gaylor, McGraw-Hill Company, pp. 4–5, 4–6.

38. Stamenkovic, H. (1980) "Special Provision Applicable to Concrete Sea Structures", discussion, *ACI Journal* 77, 124–125.

39. Craemer, H. (1949) *Theory of Plasticity of Reinforced Concrete*, edition of *Scientific Building Book*, Beograd, pp. 25–26, 33.

40. Almen, J. O. and Black, P. H. (1963) *Residual Stresses and Fatigue in Metals*, McGraw-Hill Book Company, New York, pp. viii, 6.

Chapter 5

A triangularly reinforced shear wall can resist higher lateral forces better than an ordinary shear wall

5.1 A brief overview of the problem

This chapter explains why Ritter and Morsch's concept for calculating web reinforcement in a shear wall is obsolete and proposes a new idea for a much safer design of a shear wall which will resist lateral forces better. This proposal is based on the concept of triangular rigidity, which is achieved by applying diagonal reinforcement in a shear wall. Further, this chapter explains the fundamental differences between a simple beam and a shear wall and shows how triangular reinforcement in a shear wall can greatly increase its resistance to lateral loads and mandates ductile failure only (due to tension). In order to increase the safety of a shear wall, it appears that the most rational design (with the highest degree of safety) would be triangular reinforcement, because external lateral forces would be applied to act parallel to the reinforcement (as occurs in a truss). Consequently, the most rational application of reinforcement would be achieved. This would also be the safest design because a triangle is the only rigid geometrical configuration, and any diagonal tension (stretching) could be controlled by corresponding diagonal reinforcement (such as a triangle's hypotenuse). This is in contrast to the existing concept that additional horizontal wall reinforcement does not contribute to shear strength.[1,2] Both types of reinforcement represent a similar financial burden; yet, triangularly placed reinforcement creates rigid triangles which are far more effective against lateral loads.

In Section 5.9, this chapter suggests new thinking on subfoundations that will absorb a large portion of the shock from an earthquake before it reaches the structure. Small sliding movements of the structure in all directions will also increase its safety.

In Section 5.10, diagonal cracking in a ductile steel frame is explained. Diagonal cracking in a column's web, at the connection with a horizontal beam, is a result of internal active compression and tensile forces combined with vertical axial shear forces, acting as resultant punch shear force V_n at a given crack.

In Section 5.11, examples of the structural calculations for a classical shear wall as per the ACI 1995 and the new triangularly reinforced shear wall are shown. The results of the testing of such walls under lab conditions are also shown.

5.2 Introduction

During the 1964 Alaskan earthquake, spectacular diagonal tension failure suggested that the coupling beams of shear walls are probably inherently brittle components.[3] Even structures totally complying with the UBC showed diagonal tension failure in such a way that no reasonable explanation emerged.[4] "Also, the relative contribution to shear strength provided by vertical and horizontal web reinforcement is not fully understood,"[2] ... "It is likely that the near future will see radical changes in design methods of resisting diagonal tension."[5]

This is exactly what this chapter does. It suggests an incomparably safer design. A corresponding desire to learn more about earthquakes could be illustrated by the following quotes: "It demands understanding of the basic factors that determine the seismic response of structures, as well as ingenuity to produce systems with the required properties;"[6] ... "Present shear and bond seismic provisions appear to be inadequate."[7] ... "Henry Degenkolb showed the (1973) code-designed Imperial County Services Building that was torn down as a total loss following a moderate southern California earthquake in 1979 because its stylish open ground floor weakened it (*Science*, 29 August 1980, p. 1,006). The old masonry courthouse across the street suffered no damage. He showed several buildings in downtown Managua after the great earthquake there. All were still standing, but the one designed to resist the greatest shaking suffered the most damage, while the one with no seismic design suffered the least damage. "Doesn't this tell us that perhaps we're on the wrong track?" Degenkolb said. "The real guts of earthquake engineering is not contained in present codes. We don't fully understand the tie-in between what we measure (severity of shaking) and damage."[8]

The addition of more stirrups, as calculated by the truss analogy theory, soon reached a stage where more stirrups would not increase resistance to diagonal cracking in a shear wall. "It is not surprising, therefore, to find from experimentation that additional stirrups did not improve shear strength."[1] ... "For the specimens with a height-to-horizontal length ratio of $\frac{1}{2}$ and less, it was found that horizontal wall reinforcement did not contribute to shear strength."[2] For this reason, because of our very limited ability to design safer shear walls, it has become necessary to overcome the deficiency of the existing shear wall design techniques.

In fact, the purpose of this work is to contribute to a higher safety factor of resistance against seismic forces by introducing a new concept of shear wall design. This is necessary because the seismic forces prescribed in the codes as static forces are usually considerably lower than those forces imposed during a major earthquake.[9] Besides, "Earthquake engineering is still an art and not a science. Seasoned engineering judgement is a critical ingredient."[10] The accuracy of the quote above and also the very limited scientific knowledge for structural design against earthquake failure can be seen from the following quote: "The relative contribution to shear strength provided by vertical and horizontal web reinforcement is not fully understood."[2] The facts given above have been the main motivation for suggesting this new concept of reinforcement in a shear wall, which will provide for safer structural design in seismically sensitive zones.

5.3   Theoretical analysis

The following brief comments will serve to avoid confusion between a cantilever as a statical member, and the so-called "cantilever shear wall" as another statical member (which is fundamentally different from the former). Notably, any beam on two supports is composed of two cantilevers, anchored to each other at the point where vertical shear forces are zero. In such a case, physical supports (reactions R_1 and R_2) are external concentrated loads, located at the ends of each cantilever, acting downward, while real external loads act upward, as illustrated in Figure 2.10. In other words, if we rotate a simply supported beam by 180°, and if we imagine that some anchoring wall passes perpendicularly through the point where vertical shear forces are zero, we will have two cantilevers. So, when we use the phrase "beam on two supports" in this study, it means a statical member composed of two cantilevers. In reality then, any deformation of any simply supported beam represents the deformation of two cantilevers, anchored (attached) to each other at the point where vertical shear forces are zero.

In that sense, the classical theory of shear wall design is based on the design theory for a cantilever (beam on two supports) instead of being based on the theory for deformation of a member exposed to pure shear action, as any shear wall generally is: a statical member exposed to the action of pure shear. For this purpose, it is necessary to introduce the following two definitions:

1   A shear wall is any wall in which, under loaded conditions, one diagonal is exposed to tension and another to compression. Such a wall will be treated as suggested in this chapter.
2   A cantilever shear wall is any shear wall in which, under loaded conditions, neither diagonal is exposed to compression or tension. This will be treated, as suggested in Section 2.6 of Chapter 2, for any beam where a neutral plane (axis) is developed.

Besides numerous other differences, the following is the primary one: in a shear wall, the permanent tendency to be stretched (elongated) on one of its diagonals exists; while in a beam on two supports, such elongation (stretching) of its diagonals does not exist. So, to visualize the main differences, because the theory of a simply supported beam is not applicable to shear walls, there will be separate discussions for a beam on two supports and for a shear wall.

A beam on two supports will be investigated from the point of view of its nature of deformation and its possible failure. It will not be investigated from the viewpoint of any assumption that uses conditional words like "if" or "maybe," or from the point of view that "someone said it to be so." For example, "The origin of the concept of diagonal tension is uncertain, but a clear explanation of diagonal tension was presented by W. Ritter as early as 1899,"[11] and "Morsch pointed out that *if state of pure shear stress exists*, then a tensile stress of equal magnitude must exist on a 45-degree plane."[11] The same rule of scrutiny will be applied to a shear wall: not from what anyone has said, but from the point of view of its nature of deformation. By so doing, we can distinguish reality from fantasy and find separate solutions for the safer design of a beam on two supports, as well as for a shear wall. This is the primary objective of this work.

5.3.1 Beam on two supports

In order to understand the phenomenon of diagonal cracking in a beam on two supports; a portion of our latest findings will be introduced. The innovation is, that in a flexural member, internal active and internal resisting forces are developed simultaneously. The combination of such internal active vertical shear forces (V) with the internal active compression forces (C) and tensile forces (T) leads to diagonal cracking and diagonal failure as illustrated in Figure 5.2(a).

It should be mentioned here that in a flexural member, Newton's third law (where any external action, as R_1, R_2 and F, causes its own internal reaction, as R_{1r}, R_{2r} and F_r, as shown in Figure 5.1) coexists with our new law, where any internal active force (C, T, H, V) causes its own internal resisting force (C_r, T_r, H_r, V_r), as illustrated in Figure 5.1.

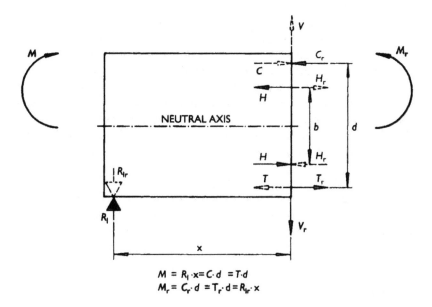

$$M = R_1 \cdot x = C \cdot d = T \cdot d$$
$$M_r = C_r \cdot d = T_r \cdot d = R_{1r} \cdot x$$

Figure 5.1 Equilibrium of a free body becomes algebraically provable by application of our new law: $\sum X = \sum Y = 0$, and $\sum M = 0$ because $C = C_r$, $T = T_r$, $V = V_r$, $H = H_r$, $R = R_r$ and $M = M_r$.

5.3.1.1   Diagonal cracking is caused by a combination of internal active shear forces with internal active compression and tensile forces

As can be seen in the author's Figure 5.2(a) (and in Figures 1(b), 1(c) and 1(e) of reference 12), the cracks shown are not caused by diagonal tension, but rather by pure punch shear: all cracks are oriented from the external forces directly towards the supports. Illustration of such law of punching, where a cracking line starts at one force and ends at another (here, at the support), can be seen in Figure D of reference 13 as well as in Figure B (on p. 154, beams 1–8) of reference 14, and in Figure 7.20 (p. 308) of reference 1. Figure B[14] is one of the finest examples of pure punch shear cracking in a bent beam, because the direction of cracking is a function of the distance between two oppositely acting punching forces.

All of the above explanations point to the fact that diagonal cracking is a result of the natural tendency of the support to move one portion of the flexurally bent member upward and the natural tendency of the external load to move another portion of the beam in the direction of its action (here, downward).

Common sense tells us that such action of a support and the external load will cause some diagonal cracking as the only possible failure of such a member. This will be the case if it previously did not fail in flexural bending, where the ultimate tensile resisting stresses are very low in comparison to ultimate compressive resisting stresses. This is the nature of concrete and any other form of masonry.

Moreover, anyone who has witnessed the pure and clear diagonal punch shear wrinkles develop in a flexurally bent thin metal double T beam, can indeed testify that these wrinkles (buckling) are caused by the tendency of a support to move its portion of the beam upward and the tendency of the external load to move its portion of the beam downward. Such opposite tendencies of movement of the same member lead to the creation of "wrinkles" as the only possible deformation caused by the movement in opposite directions. If the same beam was made of concrete, such "wrinkles" would be manifested as diagonal cracking caused by punch shear forces, precisely as shown in Figure 5.2(a). The tendency of such movement and the result of such movement, manifested as "wrinkles" or eventual diagonal cracking, is the best proof of the ever-present punch shear in a flexurally bent member.

Figure 5.2(a) clearly illustrates that diagonal cracking in a bent member is caused not by any diagonal tension but by pure resultant punch shear force V_n, acting perpendicular to any diagonal crack. Such forces V_n are the combination of internal active compression forces (C) and tensile forces (T) with the internal active forces of the vertical shear (V).

5.3.2   Shear wall

Unfortunately, even though any shear wall looks like a cantilever, its behavior during deformation and its stress condition do not have the slightest resemblance

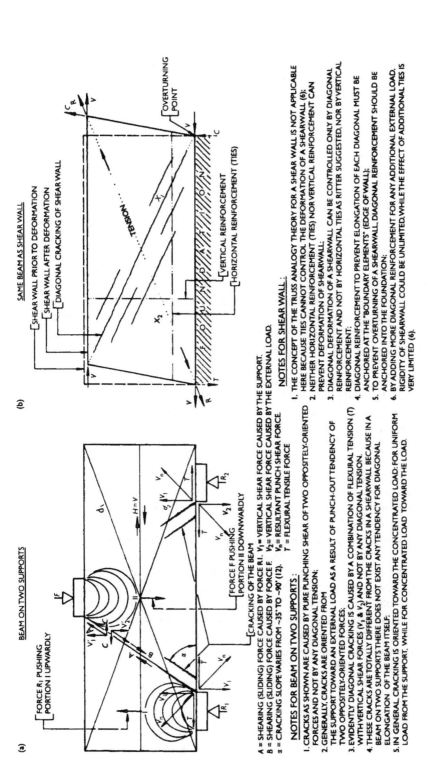

(a)

BEAM ON TWO SUPPORTS

FORCE R₁ PUSHING PORTION I UPWARDLY

FORCE F PUSHING PORTION II DOWNWARDLY

CRACKING OF THE BEAM

A = SHEARING (SLIDING) FORCE CAUSED BY FORCE R1. V_1 = VERTICAL SHEAR FORCE CAUSED BY THE SUPPORT.
B = SHEARING (SLIDING) FORCE CAUSED BY FORCE F. V_2 = VERTICAL SHEAR FORCE CAUSED BY THE EXTERNAL LOAD.
α = CRACKING SLOPE VARIES FROM –35° TO –90° (12). V_n = RESULTANT PUNCH SHEAR FORCE
T = FLEXURAL TENSILE FORCE

NOTES FOR BEAM ON TWO SUPPORTS :

1. CRACKS AS SHOWN ARE CAUSED BY PURE PUNCHING (SLIDING) FORCE OF TWO OPPOSITELY-ORIENTED FORCES AND NOT BY ANY DIAGONAL TENSION;

2. GENERALLY, CRACKS ARE ORIENTED FROM THE SUPPORT TOWARD AN EXTERNAL LOAD AS A RESULT OF PUNCH-OUT TENDENCY OF TWO OPPOSITELY-ORIENTED FORCES;

3. EVIDENTLY DIAGONAL CRACKING IS CAUSED BY A COMBINATION OF FLEXURAL TENSION (T) WITH VERTICAL SHEAR FORCES (V_1 & V_2) AND NOT BY ANY DIAGONAL TENSION.

4. THESE CRACKS ARE TOTALLY DIFFERENT FROM THE CRACKS IN A SHEARWALL BECAUSE IN A BEAM ON TWO SUPPORTS THERE DOES NOT EXIST ANY TENDENCY FOR DIAGONAL ELONGATION OF THE BEAM ITSELF;

5. IN GENERAL, CRACKING IS ORIENTED TOWARD THE CONCENTRATED LOAD; FOR UNIFORM LOAD FROM THE SUPPORT, WHILE FOR CONCENTRATED LOAD TOWARD THE LOAD.

(b)

SAME BEAM AS SHEAR WALL

SHEAR WALL PRIOR TO DEFORMATION

SHEAR WALL AFTER DEFORMATION

DIAGONAL CRACKING OF SHEAR WALL

TENSION

OVERTURNING POINT

VERTICAL REINFORCEMENT

HORIZONTAL REINFORCEMENT (TIES)

FOUNDATION

NOTES FOR SHEAR WALL :

1. THE CONCEPT OF THE TRUSS ANALOGY THEORY FOR A SHEAR WALL IS NOT APPLICABLE HERE BECAUSE TIES CANNOT CONTROL THE DEFORMATION OF A SHEARWALL (6);

2. NEITHER HORIZONTAL REINFORCEMENT (TIES) NOR VERTICAL REINFORCEMENT CAN PREVENT DEFORMATION OF SHEARWALL;

3. DIAGONAL DEFORMATION OF A SHEARWALL CAN BE CONTROLLED ONLY BY DIAGONAL REINFORCEMENT AND NOT BY HORIZONTAL TIES AS RITTER SUGGESTED, NOR BY VERTICAL REINFORCEMENT;

4. DIAGONAL REINFORCEMENT TO PREVENT ELONGATION OF EACH DIAGONAL MUST BE ANCHORED AT THE 'BOUNDARY ELEMENTS' (EDGE OF WALL);

5. TO PREVENT OVERTURNING OF A SHEARWALL DIAGONAL REINFORCEMENT SHOULD BE ANCHORED INTO THE FOUNDATION;

6. BY ADDING MORE DIAGONAL REINFORCEMENT FOR ANY ADDITIONAL EXTERNAL LOAD, RIGIDITY OF SHEARWALL COULD BE UNLIMITED, WHILE THE EFFECT OF ADDITIONAL TIES IS VERY LIMITED (6).

Figure 5.2 Here (a) shows typical cracking and failure of deep beam caused by pure punching shear; while (b) shows typical cracking and failure of shear wall caused by pure diagonal tension.

to the behavior and stress condition of any flexural member. In fact, its behavior and its stress condition illustrate a member exposed to pure shear. Yet, presently a shear wall is designed as a flexural member.

During a discussion on the concept of pure shear, Timoshenko, in discussing his Figures 3.11(c) and 3.12 stated: "To visualize this stress condition more readily, we rotate the element abcd by 45° and place the edges b'd' and bd in juxtaposition as shown in Figure 3.12."[15] The shear wall in Figure 5.2(b) is a carbon copy of Timoshenko's Figure 3.12. This means that we are literally speaking about the deformation of a shear wall as deformation due to pure shear and not of any cantilever as a part of a flexural member.

We can call the shear wall in Figure 5.2(b) a "cantilever," "a portion of a flexural member" or any other name we like to choose, the bottom line is that all shear walls, in general, are exposed to pure shear. Such figures have no correlation with the deformation of a flexurally bent member. In a shear wall, one diagonal is exposed to stretching (tension) and another to shortening (compression). In a real cantilever, as a part of a simply supported beam, such deformation of diagonals does not exist. Also, a beam on two supports (as a member of a building assembly) is never loaded in such a way as to resemble an element exposed to pure shear, where one diagonal will be elongated and the other shortened, as is the case with the shear wall in Figure 5.2(b).

In a shear wall, shear strain (elongation of a given unit) could be measured directly and, by multiplying such strain with its modulus of elasticity, stresses are determined. Yet, if a shear wall were a real cantilever, why did anyone not ever succeed in directly measuring such shear strain in a beam on two supports? Instead, the existence of diagonal tension in a beam on two supports is determined indirectly by conversion (due to trigonometry) of principal stress (obtained by multiplying principal strain with its modulus of elasticity) into shear stress. The so-determined stress appears to be oppositely oriented to real shear sliding stresses.[16] Such conversion is a simple mathematical game and has nothing to do with the real stress condition of a bent member.

Consequently, to avoid showing any example of successfully measured shear in a beam on two supports, Timoshenko made the following statement: "However, accurate measurement of a shearing strain is found to be very difficult. It is easier and more accurate to measure the principal strains"[15] and using trigonometric formulas (Mohr's circle), convert such stresses into shear stresses. Yet, horizontal sliding along a neutral plane (axis) never follows such direction, as shown on any unit element; rather sliding appears to be oppositely oriented.[16]

Finally, if any shear wall is a "cantilever" as a portion of a beam on two supports, then Timoshenko's Figure 3.12,[15] or any other figure of pure shear, becomes a "cantilever" and the case of pure shear would disappear in favor of the so-called "cantilever."

The necessity of shear walls for future structures could be illustrated by the following quotes: "Actually from observation in earthquakes, it seems that we can no longer afford to build our multi-story buildings without shear walls,"[17] ... "The use of shear walls or shear equivalents becomes imperative in certain high-rise buildings if inter-story deflections, caused by lateral loading, are to be controlled. Not only can they provide adequate structural safety, but they also give a great measure of protection against costly non-structural damage during moderate seismic destructions."[1]

Any shear wall provides much more lateral stiffness than a moment-resisting frame and by triangular (truss) reinforcement, its ductility will surpass that of a moment-resisting frame as can be seen in Section 5.10.1. of this chapter. This is because any diagonal failure will be controlled by truss reinforcement and not by the concrete itself.

5.3.2.1 Description, main characteristics and advantages of our shear wall

A shear wall is a wall designed to resist lateral forces parallel to the wall; so for the safety of a building, such walls must exist in both directions. The essential concept of a shear wall is "rigidity," or prevention of diagonal elongation within the shear wall. Under current shear wall design practices, horizontal stirrups are present to prevent abrupt failure (even though "additional stirrups did not improve shear strength").[1] But this does not satisfy the main purpose of a shear wall, which is to control rigidity and to prevent diagonal failure. In contrast, triangular reinforcement in a shear wall, as proposed herein, will accomplish these two goals of rigidity and prevention of diagonal failure, or punching failure, due to concrete.

A sudden failure mode of a shear wall due to shear must be suppressed if structural survival in large seismic disturbances is to be achieved: "In earthquake-resistant structures in particular, heavy emphasis is placed on ductility and, for that reason, the designer must ensure that a shear failure can never occur."[1] By preventing sudden failure of a shear wall due to shear, the ductility of such a member due to flexural bending is simultaneously achieved, and thus the only possible failure is tension failure.

Applying the truss concept by placing stronger vertical reinforcement at the edges and stronger horizontal reinforcement at the top and bottom of the wall, and with diagonal reinforcement, the maximum possible rigidity will be achieved by the formation of four rigid triangles connecting the top and bottom of the wall. Diagonal tensile forces can be totally controlled by such diagonal reinforcement, and by its anchorage into the foundation, it will transfer such forces into the ground. With such diagonal reinforcement, the total required reinforcement (including horizontal and vertical) could be the same or less than that required by the classical concept but with a much higher resistance to failure.

Generally, two main modes of shear wall failure have been observed in experiments performed in several countries:

a   Diagonal cracking starting at a vertical edge of the wall in the flexural tensile zone and propagating diagonally therefrom;

b   If there is sufficient reinforcement at the vertical edges of the wall to prevent diagonal cracking there, then the cracking will start from the top of the wall and extend towards the overturning point of the wall. This is particularly true of square or squat walls.[26]

Based on the facts noted above, especially mode (b), which is fully illustrated in reference 18, it appears necessary to put horizontal reinforcing in the top of a shear wall to prevent formation of diagonal cracking there. There is no alternative solution to this method. Consequently, by applying reinforcement triangularly, as required for any simple truss, a better shear wall will be achieved with a smaller amount of reinforcing (as in the conventional theory of design as a cantilever) and with much higher safety. Thus, the structural analysis of any shear wall, made from reinforced concrete (RC) members, will be governed by existing engineering laws for structural truss analysis.

It should be separately emphasized that the existing concept of shear wall design does not allow full exploitation of reinforcement ("additional stirrups did not improve shear strength").[2] The suggested truss reinforcement (triangular) allows total exploitation of reinforcement in compression and tension, which indicates that the most rational and the most economical design with the highest degree of safety will be achieved.

The point is that in any truss, members start to resist corresponding shear stresses as soon as a load is imposed. These shear forces would be under control by the proposed new diagonal reinforcement acting longitudinally through any bar as soon as such forces are developed in the member.

In a classical shear wall design, having horizontal and vertical reinforcement, only that reinforcement crossed by corresponding concrete cracking will resist the applied shear forces while all other reinforcement remains neutral, or idle.

The main point of this discussion is that, in classical design, only reinforcement crossed by concrete cracking will cooperate in resisting applied shear forces while all other reinforcement is literally neutral, or idle.[1,2] In this new design, all reinforcement simultaneously cooperates in resisting corresponding shear forces. Consequently, shear cracks in the concrete can appear only after the reinforcement itself reaches its own yield point while in the classical design, wall cracks begin to appear as soon as the tensile yield strength of the concrete is reached.

As illustrated in Figure 5.3(c), a truss can be designed to handle additional stories if the forces of the additional stories act as flexural forces on the truss members. Consequently, such a member could be designed to simultaneously take axial loading as well as flexural bending using the new method to control shear forces.

From the above discussion, it is evident that arguments in favor of the classical concept of designing a shear wall as a cantilever beam are untenable. It is also evident that it will be significantly more rational, and safer, to design a truss as a cantilever instead of designing a beam as a cantilever for the reasons mentioned above.

The most important point is that this new design will allow the wall to act as a truss while resisting deformation under wind or earthquake loading, and simultaneously act as a ductile flexural member during its mode of failure.

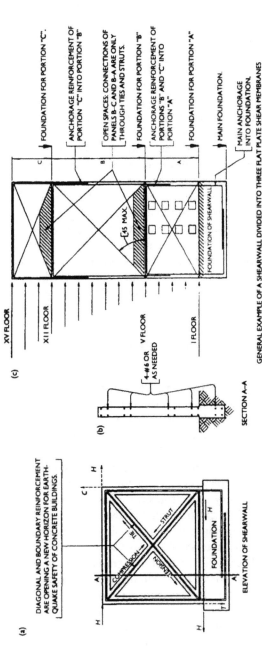

(a)

DIAGONAL AND BOUNDARY REINFORCEMENT ARE OPENING A NEW HORIZON FOR EARTH-QUAKE SAFETY OF CONCRETE BUILDINGS.

STRUT

TIE

COMPRESSION

TENSION

FOUNDATION

ELEVATION OF SHEARWALL

(b)

4-#6 OR AS NEEDED

SECTION A-A

(c)

XV FLOOR

XII FLOOR

V FLOOR

I FLOOR

≤ 15 MAX.

FOUNDATION OF SHEARWALL

FOUNDATION FOR PORTION "C".

ANCHORAGE REINFORCEMENT OF PORTION "C" INTO PORTION "B"

OPEN SPACES: CONNECTIONS OF PANELS B–C AND B–A ARE ONLY THROUGH TIES AND STRUTS.

FOUNDATION FOR PORTION "B"

ANCHORAGE REINFORCEMENT OF PORTIONS "B" AND "C" INTO PORTION "A"

FOUNDATION FOR PORTION "A"

MAIN FOUNDATION.

MAIN ANCHORAGE INTO FOUNDATION.

GENERAL EXAMPLE OF A SHEARWALL DIVIDED INTO THREE FLAT PLATE SHEAR MEMBRANES ANCHORED INTO EACH OTHER WITH POSSIBLE PERFORATION OF SUCH PLATE WITH OPENINGS.

NOTES:

1. THE CLASSICAL CONCEPT OF TRUSS ANALOGY THEORY HAS A VERY LIMITED ABILITY TO INSURE THE SAFETY OF SHEARWALLS. "IT IS NOT SURPRISING, THEREFORE, TO FIND FROM EXPERIMENTS THAT ADDITIONAL STIRRUPS DID NOT IMPROVE SHEARWALL STRENGTH"[1];

2. THE WRITER'S SUGGESTED CONCEPT, THEORETICALLY, DOES HAVE AN UNLIMITED POSSIBILITY FOR HIGHER RESISTANCE TO SHEAR FAILURES. FOR ANY INCREASES IN SHEAR FORCES THERE COULD BE AN INCREASE, TO INFINITY, OF TRIANGULAR REINFORCEMENT;

3. THE CONCEPT OF "DEEP BEAM" IS NOT APPLICABLE FOR SHEARWALLS BECAUSE THE REAL LOAD OF A DEEP BEAM AND THE REAL LOAD OF A SHEARWALL ARE FUNDAMENTALLY DIFFERENT. A DEEP BEAM IS NOT EXPOSED TO ANY DIAGONAL STRETCHING, WHILE ANY TYPE OF A SHEARWALL IS EXCLUSIVELY EXPOSED TO DIAGONAL STRETCHING AND DIAGONAL ELONGATION;

4. THE STALEMATE OF PROGRESS IN GREATER SAFETY DESIGN OF SHEARWALLS IS DUE EXCLUSIVELY TO THE FOLLOWING FACTORS: a) THE FALSE CONCEPT OF THE RITTER–MÖRSCH TRUSS ANALOGY THEORY[20], AND b) THE FALLACY OF THE RITTER–MÖRSCH CONCEPT OF DIAGONAL TENSION (14–FIG. 8, 28).

5. IN ORDER TO REACT AS A TRUSS, PANEL "B" WILL BE SEPARATED FROM PANEL "A" BY AN OPEN SPACE; PHYSICAL CONTACT WILL BE ONLY THROUGH TIES AND STRUTS. THE SAME IS TRUE FOR PANEL "C".

Figure 5.3 Here (a) shows a new concept of shear wall rigidity achieved by triangular reinforcement of a flat plate concrete member where diagonal forces are controlled by diagonal reinforcement; (b) shows a cross section of such a wall with "boundary reinforcement" at the top and bottom, diagonal reinforcement and foundation reinforcement; (c) shows that the angles of diagonal reinforcement can vary as long as rigidity of the triangles is achieved; also shows the presence of vertical reinforcement to prevent sliding of a flat plate against the one below and possible accommodation of openings in such a wall.

Evidently, tensile failure would then be the only possible failure mode, because diagonal failure would be prevented by the diagonal reinforcement, which acts as a diagonal tie or strut of a truss.

It should be noted that a shear wall, where the ratio of depth (foundation) to length (height) is smaller than 1, should be treated as a beam or cantilever wall because the neutral axis will be developed while diagonals will not be exposed to elongation and shortening. In such a case, triangular reinforcement will not contribute to higher safety because the truss will not react to external load as a truss, but rather as a beam since a neutral axis has been created. However, if we decide to treat any beam (cantilever or not) as a triangularly reinforced member (or truss), then any member of the truss (ties, struts, columns) should be cast in concrete with free space between the trusses' members, and such member, under loaded condition, would react as a truss and not as a beam because the neutral plane (axis) will not be developed. In other words, any beam could be replaced with a truss, if economy required, as illustrated in Section 5.11.6.2.

A shear wall where the ratio of depth (foundation) to length (height) is larger than 1, should be treated as a cantilever truss because a neutral axis will not be developed, while diagonals will be exposed to elongation and shortening.

The key words here are 45° of stress distribution of any force acting perpendicularly at any plane where square (1:1) will be borderline for creation of either a shear wall or a beam.

5.4 Comparison of cracks in a reinforced concrete beam and a shear wall

Diagonal cracks in a simple beam are not and cannot be parallel, as Ritter suggested. Rather, they are always oriented on a diagonal from the supports upward, toward the external load on the top of the beam: "At the interior support of a beam, the diagonal cracks, instead of being parallel, tend to radiate from the compressed zone at the load point,"[1] ... "In most laboratory tests, if dead load is neglected, the shear span is the distance from a simple support to the closest concentrated load."[2] These concepts are exactly as illustrated in Figure 5.2(a). Such cracking is caused by pure punch shear; the tendency of the support to move its portion of the beam upward and the tendency of the external load to move its portion of the beam downward. This cracking has nothing to do with any diagonal tension, as Figure 5.2(a) clearly proves. The last quote should always be kept in mind when diagonal failure is in question because it is one of a dozen unmistakable proofs of the fallacy of diagonal tension. Forces located in diagonals d_1 and d_2 (Figure 5.2(a)) do not exist and such a crack (following these diagonals) has never been developed in a simple beam.

Cracks in any shear wall are fundamentally different from cracks in a simple beam. Shear walls fail diagonally as a result of the stretching of one of the two diagonals (x_1 or x_2 in Figure 5.2(b)), as is the case with pure shear. The tendency of one portion to move upward and another downward (as is the case with a beam on two supports) does not exist here! So, to prevent the elongation of either diagonal, and consequently to prevent cracking parallel to the other diagonal, it is evident that diagonal reinforcement as the hypotenuse of a triangle must be present because the tensile strength of concrete is almost nonexistent. On the other hand, during structural design, concrete will be allowed to participate with its corresponding stresses. To prevent pullout, diagonal reinforcement must be anchored into the bottom and the top of the wall making real triangular reinforcement. By adding special reinforcement at the edges of the wall, as is required for any truss, and by anchoring diagonal reinforcement into the boundary reinforcement, as shown in Figure 5.3, complete and true rigidity of the shear wall will be created by the formation of statically rigid triangles. In fact, four rigid triangles will be created, each bounded by diagonals and two boundary elements. Any additional bars of a triangle's reinforcement will lead to additional rigidity, or closer control of the elongation of any diagonal of the shear wall, thus providing additional resistance to diagonal cracking. In contrast, additional stirrups, as calculated by the existing theory, soon reach a stage where more stirrups would not increase resistance to diagonal cracking in a shear wall.[1,2]

By using triangular reinforcement, the concentration of axial forces will be greatly reduced in the compressed zone by the proportional distribution of such forces in a given truss, created by diagonal reinforcement. At the same time, the recognized problem, "That axial compression will reduce ductility,"[1] will almost be eliminated by the proportional distribution of such force in all truss members. Also, by allowing a proportional distribution of a given load to any member of the truss, excessive reinforcement of the boundary element will not be necessary! Consequently, the so-called "boundary element," with a much larger cross section than the rest of the wall, will be eliminated completely. Also, diagonal reinforcement of a shear wall will greatly decrease the overall displacement of a building. This element plays a big role in the psychological demeanor of the occupants in such a structure.

From research worldwide, it can generally be seen that by increasing the resistance of a shear wall to lateral loads, all desirable characteristics will be increased: "More recent studies have indicated that, with controlled flexural capacity and full protection against diagonal tension failure, considerable ductility can be extracted even from squat shear walls."[19] Common sense leads us to this natural phenomenon!

Naturally, to prevent the overturning of a shear wall and diminished concentration of compression forces at the overturning point (caused by dynamic impact to the shear wall at one point due to the vibrating forces of an earthquake), diagonal reinforcement must be anchored solidly into the foundation so that the entire structure, including the foundation, would vibrate as one unit.

Also, the concentration of reinforcement in compression zones, due to the formation of struts, will simultaneously increase ductility. This is illustrated by the following quote: "The wall with reinforcement at the ends yielded at a higher load and had the same ultimate strength but more ductility than the other wall (with an equal amount of reinforcement) even though it has less vertical reinforcement."[20]

Furthermore, the bending moment, shear forces caused by lateral loads and axial compression caused by gravity loads, will be completely controlled by rigid trusses formed by the diagonal reinforcement of shear walls. Naturally, any compression strut should be provided with transverse ties around the bars to prevent their buckling and to retain the cracked concrete core within the compressed bars at the extremities of the strut section.

It becomes very clear that the existing concept of the design of shear walls has a very limited possibility for increasing safety of such a wall,[1,2] while the concept of rigid triangles has an almost unlimited potential to increase resistance to diagonal failure. This is because the function of such a wall is related to the amount of diagonal reinforcement versus the diagonal elongation of the wall itself.

We emphasize here that the rigidity of triangular (truss) reinforcement in a shear wall should be clearly distinguished from the possible diagonal bracing done by Gallegos and Rios as a method of repair of earthquake damaged structures.[21] The concept of bracing is truss, but has nothing to do with the truss concept of reinforcement in a shear wall.

Because the foundation and walls (applied in both directions) will react as one structural element and vibrate in the same mode, the dynamic impact will be transferred directly to the ground. Such "soft" ground, as suggested in Section 5.9 will serve as an excellent damping base (simulating a ship floating on a stormy sea) if the RC structure starts behaving as a rigid box due to the rigidity of its shear walls. Also, the resistance of a shear wall to possible failure could become unlimited, because unlimited addition of triangular reinforcement is possible!

5.5   Critical review

In a shear wall it is not possible to ever have the ultimate shear resistance (caused by horizontal and vertical reinforcement) to be equal to the ultimate shear resistance (caused by the triangular reinforcement) in a second (new) shear wall, even if they are exposed to the same lateral forces and the same quality of concrete with the same quality and quantity of reinforcement. The reason is a very simple one: rectangular reinforcement, exposed to diagonal tension, has no natural resistance to deformation of a given configuration abcd (Figures 5.4(a) and (b)). Resistance is only achieved by the concrete itself and not by reinforcement. This is because the resultant force D is free to cause cracking through diagonal d–b without any participation of a given reinforcement. In other words, the configuration in Figure 5.4(a), after exhaustion of the ultimate concrete resistance to diagonal tension, will become the form shown in Figure 5.4(b), with negligible contribution of reinforcement to diagonal cracking.

In conclusion, it could be said that the resistance to deformation of the configuration abcd (as shown in Figure 5.4(a)) is exclusively the function of rigidity

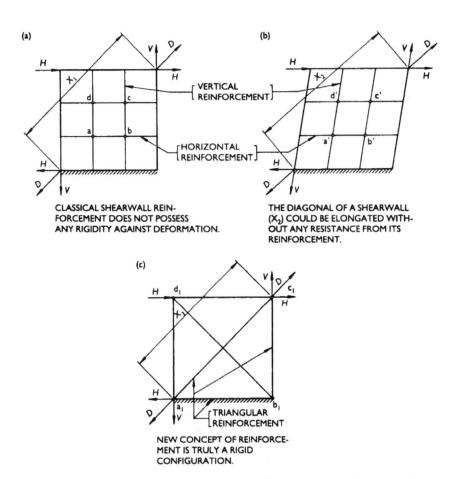

(a) CLASSICAL SHEARWALL REIN-
FORCEMENT DOES NOT POSSESS
ANY RIGIDITY AGAINST DEFORMATION.

(b) THE DIAGONAL OF A SHEARWALL
(X_2) COULD BE ELONGATED WITH-
OUT ANY RESISTANCE FROM ITS
REINFORCEMENT.

(c) NEW CONCEPT OF REINFORCE-
MENT IS TRULY A RIGID
CONFIGURATION.

Figure 5.4 Shear wall reinforcement as per the classical theory (as required for a cantilever)
does not offer any rigidity against elongation of its diagonals. In (a), the rectangle
abcd, under the smallest force, will change configuration into rhomboid a'b'c'd'
as shown in (b). In this case, rigidity is a function exclusively of concrete strength,
so any addition of horizontal bars (stirrups or ties) cannot increase the resistance to elongation of diagonal x_1, consequently diagonal cracking will take place
after exhaustion of the shear strength in concrete. As a result, diagonal x_1 will
be transformed into a longer diagonal x_2, without any contribution of reinforcement to prevent such elongation! With substitution of triangular reinforcement
in (c) (contrary to (a) and (b)), elongation of diagonal x_3 is possible only after
reaching the plastic stage of the reinforcement where any bar, together with
concrete, participates proportionally in sharing the imposed lateral load.

of the concrete itself and not of its reinforcement. For that very simple reason,
any additional reinforcement could not increase the resistance to shear failure, as
reported by Barda–Hanson and Corley 1978.[2] Yet, if a shear wall with triangular
reinforcement is exposed to diagonal force *D* (as shown in Figure 5.4(c)), then the

following four facts become evident:

1   The configuration shown is a totally rigid one, since its deformation is a function of diagonal reinforcement.
2   The entire force D is proportionally distributed between the compression strut, the tensile tie and the corresponding chords, as in any truss.
3   The total exploitation of reinforcement is possible because forces act parallel to the bars, and the deformation for causing cracking is possible only after reaching its plastic stage.
4   Any additional reinforcement will increase the resistance to the diagonal cracking of a given wall for its ultimate capacity to remain in an elastic stage.

The fact should be clearly emphasized that triangular reinforcement has no correlation to squat walls, where Park and Paulay[1] applied diagonal reinforcement between two windows. The differences from the technical or economical points of view are not comparable, as can be seen from the following analysis:

1   Park and Paulay used diagonal reinforcement (not triangular) only for a short beam (coupling beam) and not (as we have suggested), for a shear wall as a whole.
2   Their anchorage represents an enormous problem: for a 3 ft long beam they must have 3.5 ft of anchorage at any side,[19] while triangular reinforcement, as a truss, does not need any anchorage except one into the foundation.
3   Their concept is based only on compression strut and tensile ties, while ours is based on the rigidity of a truss itself.
4   With our concept, all such diagonally reinforced coupling beams would be eliminated by saving the entire reinforcement which was necessary for performance of their roles.
5   Reinforcement, as determined for a shear wall by calculating it as a cantilever, will be more than enough to satisfy the need of our trusses for the same wall, thus giving much higher resistance to deformation and possible diagonal cracking of the same wall.
6   The safety of Park and Paulay's shear wall is based on horizontal and vertical reinforcement, where any additional reinforcement cannot increase the resistance to diagonal cracking,[2] while in our case, any additional reinforcement will proportionally increase the resistance of the shear wall.
7   As far as we know, trusses of steel, as additional shear membranes to a concrete wall, have been used; however, trusses as concrete reinforcement, without any conventional reinforcement in such a concrete wall, have not been suggested by anyone to date.
8   This is the first time it has been suggested that a shear wall be calculated as a truss and treated as a membrane exposed to pure shear and not as a cantilever beam, because a shear wall does not, in any way, behave like a cantilever beam.

9 There is no doubt that this is the first time when the existence of internal active and internal resisting forces in a cantilever have been recognized (flexural member) and that such forces are not and cannot be developed in a shear wall. Park and Paulay[1] did not recognize the existence of such forces.

10 In our case, the shear wall is treated to control the elongation of its diagonals, while elongation of the diagonals of a real cantilever beam does not even exist. This is probably the main difference between these two types of shear wall approaches to achieve much higher resistance to lateral forces.

Logically, real diagonal tension cracking in a shear wall must be oriented diagonally. Such logic has been proved in the 1971 San Fernando earthquake in California. In any shear wall where the rigidity of the wall was able to develop its mechanism of real diagonal tension, diagonal cracks developed.[22] Also, any shear wall which exhibits punch shear cracking (or horizontal cracking), as shown in Figure 1 of reference 12, unmistakably demonstrates that the main goal of such a wall – rigidity – was not achieved (for which only the designer is responsible).

Following the discussion above, it becomes obvious that triangular reinforcement for any concrete or masonry shear wall should become the main tool in designing such a wall, which will be able to provide maximum resistance to distortion due to seismic forces. This will lead to maximum safety of structures in seismically sensitive zones.

A shear wall, which interacts with the trusses in its body, will show rigidity of the trusses and will be able to control much higher lateral forces due to limited elongation of the diagonals and proportional load distribution between the chords. Failure of a diagonally reinforced wall will be governed by the yielding of tension reinforcement located in the vertical struts (ties) at the edges of the wall. Such failure is stipulated by the frame's reinforcement at the edges of the entire wall, forcing the wall to behave as a cantilever unit at the yielding stage.

An advantageously distributed reinforcement would possess the properties desired in the earthquake-resisting structures. With diagonally reinforced shear walls, overall lateral deflection and inter-story drift will be controlled. As a result of a triangularly reinforced shear wall, all portions of the membrane between ties and struts could be eliminated without any risk of shear wall failure. Consequently, the limited ductility of an ordinary shear wall, caused by coupled shear walls (short beams or columns between openings), will be totally eliminated by triangularly reinforced shear walls because of their location between struts and ties. Moreover, with the triangularly reinforced shear wall and its inherent ductility, the safety of the structure itself will be fully controlled, and the enormous damage to office or apartment buildings will be greatly diminished, if not completely eliminated. It is understandable that damages would be minimal if the entire building, with its rigid shear walls and foundation, started to vibrate as one unit or a box. During seismic vibrations, damage to the building will occur only if the structure cannot react as one rigid unit and, instead, reacts as an unstable box.

A text on soil mechanics states, "It may be assumed that the load spreads out in the underground in the form of a truncated pyramid whose sides slope at an angle of 45°."[23] This concept can also be noted in reference 14. Using common sense, it becomes evident that the stress distribution of a vertical load at the horizontal support is generally at 45°.

Following a period of over three decades of study by this author on the mechanism and control of cracking of RC members,[24] it appears to be correct that all punching shear cracks in concrete are developed, generally, at 45° if there is no interference from an additional force with its stress distribution.[13,25,26] Also, looking at the cracks in deep horizontal cantilevers,[1] as well as in a vertical shear wall considered as a cantilever,[12] and Figures 12 and 23 of reference 27, it becomes evident that the cracking follows a straight line between the external force (horizontal of a shear wall) and the support (here the overturning point or hinge point) up to 45°. When the angle of possible cracking in a shear wall exceeds 45°, that is, when the compressive stress distribution of a support and the compressive stress distribution of the external forces cannot influence each other,[14] then the direction of cracks towards an overturning point (support) will eventually change and become more erratic.

5.6   Reinforcement for our shear wall

We have discussed reinforcement in a simple and understandable manner: the entire shear wall is reinforced by continuous reinforcement, applying only one rod (wire), and installing two or four bars at any cross section of any chord of the truss. Such reinforcement will be bent in a shop or in the field, according to Figure 5.6. Before any casting of concrete, one bolt will be installed at any corner around which reinforcement is bent (see Figures 5.5–5.7).

Such reinforcement could fail only after reaching its own plastic stage and there could never be a problem of the so-called bad detailing, as is usually the case in any RC failure.

The final product of reinforcement is illustrated in Figures 5.6 and 5.7, where four bars at any cross section of the truss chords are developed from a single bar (wire).

5.7   Greater liberty for openings (windows, doors) in our shear wall

The criticism of some architects that possible openings in a shear wall could be jeopardized by such diagonal struts is unfounded for the following reasons:

1   When an architect learns that the entire area between struts and ties can be used for openings, the floor plans are drawn to accommodate the diagonal struts, as shown in Figure 5.3(c).

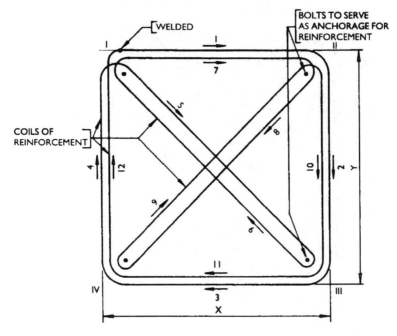

Figure 5.5 Graphical guide for reinforcing a shear wall by a single wire (bar); start with corner I and follow the pattern (arrows) from bar I to bar 12. Bars 7 and the last bar, 12 or 24 (depending on the number of layers), should be welded against each other, or against bolt I.

2   The location of diagonal struts and ties can vary at any angle as long as they satisfy the required rigidity of a given shear wall.

3   If the safety of a concrete structure can be increased greatly (in contrast to the existing concept) by designing a rigid shear wall, then openings in such a wall become of secondary importance: they can be located anywhere between truss chords.

5.8   Remarks

1   Because fundamental differences exist between a shear wall and a simple beam, fundamentally different remedies should also exist for the prevention of their diagonal failures:

   a   In a shear wall, the lateral load has a tendency to stretch one diagonal (make it longer), while in a simple beam such stretching (elongation) of diagonals does not exist.

Figure 5.6 Vertical position of reinforcement of shear wall.

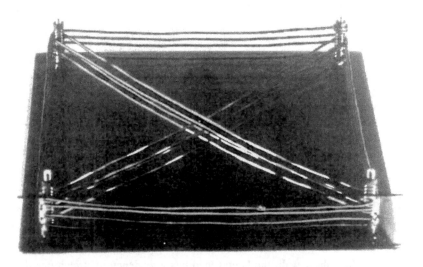

Figure 5.7 Horizontal position of reinforced shear wall with both forms.

b   Diagonal cracking in a simple beam is caused by punch shear, or by the tendency of the support to move some portion of the beam in its direction and by the opposite tendency of the external load to move its portion of the beam in its direction of action.

c   Diagonal cracking in a shear wall is caused by pure diagonal tension induced by the lateral load.

d   Web reinforcement in a simple beam (stirrups) can prevent propagation of diagonal cracking caused by punch shear, but horizontal reinforcement (stirrups) in a shear wall cannot help because horizontal and vertical reinforcement do not create a rigid geometrical configuration capable of withstanding (resisting) elongation of a given diagonal and controlling forces acting through the diagonals.

e   By placing diagonal reinforcement in a shear wall, any possible lateral loads can be controlled and a shear wall can be prevented from diagonally failing in a shear.

f   Triangular reinforcement will greatly decrease the displacement of a structure, minimizing the fear of its tenants for their safety.

2   This is the first time diagonal tension is controlled in such a way that shear forces must act longitudinally in any reinforced bar; in the classical method, shear forces act transversely in any reinforced bar.

3   Triangular reinforcement of a shear wall will reduce the concentration of axial forces in the vertical boundary zone and eliminate the necessity for a special boundary element of the shear wall.

4   Reinforcement of a simply supported beam along the diagonals cannot prevent diagonal cracking and diagonal failure.

5   Because a diagonally reinforced shear wall could become immune to diagonal failure due to concrete, it would be forced to fail only due to reinforcement, ensuring ductile failure, which is a desirable phenomenon in seismically sensitive zones.

6   Because triangular truss reinforcement allows total exploitation of reinforcement, it follows that such a design will be the most rational and the most economical.

7   It becomes clear that the existing concept of shear wall design has a very limited possibility for increasing the safety of the wall. Conversely, unlimited increase of resistance, to diagonal failure, is possible with the concept of rigid triangles, because the function of such a wall is related to the amount of triangular reinforcement against the diagonal elongation of the wall itself.

8   In contrast to an ordinary shear wall, openings between chords in our shear wall have very little, if any, influence on the resistance to lateral loads.

9   The new subfoundation concept suggests that it will absorb the largest portion of shock and cause soft vibration of a rigid structure with very little movement in any direction. This will lead to much greater safety of the structure itself.

5.9 Subfoundation

A "soft" subfoundation is essential to minimize the shock to a building caused by the sudden dynamic impact of an earthquake. Such soft subfoundation should serve as a cushion between the rigid structure (box) and the supporting ground. This can be accomplished by placing a couple layers of sand and clay, isolated (separated) by prefabricated flexible plastic material (rubber-type), simulating, as a whole, a flexible plastic membrane:

1   first layer of prefabricated flexible plastic, approximately 5–10 in thick;
2   second layer of fine sand, approximately 10–20 in thick;
3   third layer, again of prefabricated flexible plastic, 5–10 in thick;
4   fourth layer of sandy clay, 10–20 in thick;
5   fifth and last layer of prefabricated flexible plastic, 5–10 in thick; and
6   the concrete foundation, rigidly fixed to rigid shear walls, will be located directly on the plastic layer. Such a foundation during a vibration will serve as an excellent shock absorber, causing soft vibration of the rigid box structure; somewhat like the concept of a floating ship on a stormy sea.

The subfoundation and foundation will be isolated on both sides by a plastic spacer (3–5 in thick) against direct contact with solid soils or rock. The higher safety of a rigid box on a softer foundation has been proved by Frank Lloyd Wright's Imperial Hotel in Tokyo where the structure withstood the Kanto earthquake (1923), while all other surrounding buildings were heavily damaged or destroyed.[10] This will also permit a structure to have a small amount of movement in all directions, enabling it to behave as a floating structure when the ground behaves as if it liquefies, leading to a much higher factor of safety for the structure itself. Consequently, a further increase of safety results when a small amount of sliding is allowed in all directions.[28]

Evidently, this will be the lowest-cost subfoundation in comparison with new metal vibrating absorbers!

5.10 Effects of internal active forces in seismic failures of ductile steel frame structures: a consequence of the new law of physics

One of the eminent members of the Society of Structural Engineers of California, Mr. Robert Englekirk, in his communication, "Design Implications Derived from the University of Texas, Austin, Test Program" (1995) stated: "The Northridge Earthquake uncovered what now must be viewed as a fundamental flaw in the previously accepted means of joining beams and columns in a steel ductile frame (Special Moment Frame – SMF) ... During the Northridge Earthquake thousands of these steel beam-to-column connections failed and even more disconcerting was the fact that there was no evidence that the beams of panel zones experienced any post-yield behavior."

In other words, this earthquake has indicated, to our profession, that the essence of the problem lies in the lack of comprehension of the shear failure mechanism, especially since the calculation of ductile frame construction was made according to Ritter–Morsch's diagonal tension theory, which is fundamentally incorrect. Therefore it is not astonishing that unexpected failures occurred. Basically, the problem lies in the lack of comprehension of the shear failure problem.

Diagonal failures or diagonal cracks in ductile steel construction cannot be rationally explained by the existing theory of diagonal tension, but only by the action of internal active compression, tensile and vertical shear forces. The vertical cantilever vibration of the structure, conditioned by the seismic forces, is vibrating horizontally as well, and the tendency appears as a lifting up and lowering down of the structure. In such horizontal vibrations, in the domain of connections of the beam with the column, the alternation of the zones of compression and tension occur as soon as axial tensile forces acting upward appear in the column, the fibers of the compression zone becoming tensile.

When the upper tensile zone of the beam fixed into the column becomes the compression zone, unexpected diagonal cracks appear in the column, oriented upward, as shown by Figure 5.8. But these cracks are found only in the web of the column, directly from the connection of the flange with the web of the column. This is the effect of the resultant force V_n, due to tensile forces in the beam and to the axial tensile forces in the column, acting upward and downward from the crack, as illustrated in Figure 5.8. Further, this implies that the columns at the connections with the beams must be strengthened in the web, by rolling or by welding a plate insert, so that the common cross section in this place could receive and absorb the foreseen resultant punch shear force V_n, that can appear by some vibration of the structure on the given seismic soil. Thus, the sudden failure through the column will be eliminated, and ductile failure will occur through the beam. The failure at the welded connection of the column and beam can be managed by strengthening the upper and lower parts of the flange. This can be accomplished in the rolling of the beam itself or by welding additional plate pieces at the top and bottom of the flange. This is done in order to remove the inflection point (PI) from the support and to allow the reception of a greater tensile force in the area of the maximal negative moment. In this way, ductile failure would occur in the proximity of the PI, which serves as the statical support of two contrarily bent beams. In other words, by applying the acquired knowledge of the causes conditioning such cracks and with assumed maximal tensile axial force in the column, we can make the connection of a steel frame immune to sudden failure through the column and secure ductile failure through the beam for a given load of the given seismic zone.

For an upward-oriented crack to appear in a column, tension must exist in the lower flange of the beam, due to the alternation of the negative moment acting downward (moment at the fixed end), while the positive moment between the two

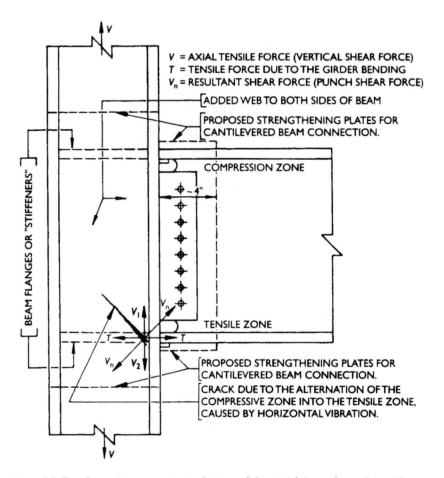

Figure 5.8 This figure illustrates the mechanism of diagonal failure of a rigid steel frame, caused by the combination of internal active tensile forces with internal active vertical shear forces, producing punch shear failure.

PIs is acting upward. This is the result of the horizontal vibration of the frame. It is evident that these cracks are not due to the diagonal tension of Ritter–Morsch, but to the combination of the internal active tensile forces T with the internal active axial forces V (acting as vertical shear forces V_1 and V_2), conditioning the cracking force V_n, acting normally to the given crack. Such a failure can easily be perceived by connecting two planks at their ends at a 90° angle, with each end being cut at a slope of 45°. The connection on these cuts is achieved by gluing that is weaker than the strength of the planks. Then, if we press the ends of these planks against each other, these planks, as a part of the frame, will separate by cracking only at the connection.

As can be seen in Figure 5.8, the crack starts in the web because the flange has a much larger cross section to resist the action of a given resultant of punch shear force V_n. This is similar to the phenomenon of diagonal cracking in a concrete web of a double T beam, where the crack starts in the web, as explained in Section 2.4.4.3 of Chapter 2 of this book.

On the other side, at the connection of the flexurally tensile flange on the upper zone of the beam, the flexural compression in the vertical flange of the column is conditioned, so that the upward-oriented diagonal crack which spreads through the compression zone of the column can never occur, as suggested by some authors in the literature (John A. Martin, Steel Moment Frame Connection, Advisory No. 3, Applied Technology Council, California, 1995, Figure C-38).

Furthermore, the resulting vibrating forces caused by an earthquake could exceed the assumed forces used in calculations so that the hinge (PI) could reach the connecting plane of the column and the beam. In such a case, with the elimination of the rigidity of a given steel frame, this plane becomes a statical support for the beam with the action of only two vertical shear forces: V, left at the column's face and V, right at the beam's face. At such a cross section (PI) there is no other force (see Section 4.5.2.1 of Chapter 4 and Figure 4.8). The previously mentioned forces only have the tendency to slide (shear) the beam vertically at the connecting plane of the column. Additionally, with the tendency to slide up and down the given beam, the resulting vibrations create the phenomenon of metal fatigue, with the vertical cracking of the beam, precisely through the area of the beam welded to the column, being an enormous contributor to the acceleration of such sliding. Some residual stresses, caused by good or bad welding, will be parallel to the vertical shear forces (V, left and V, right) acting on the same planes. Such stresses contribute to the development and propagation of vertical cracks at the connecting cross section of the beam and column, as shown in Section 4.5.4. of Chapter 4 of this book.

All this suggests that the PI or hinge must be precisely predetermined by designing its location to prevent the PI from ever reaching the connecting plane of the beam to the column. This could be achieved by making smaller cross sections of the beam at some distance from the column. Such design is of the utmost importance in guarantying the safety of the rigid beam. The location of the PI should be foreseen by the designer, as suggested in Figure 5.8, or one could simply cut a portion of the flanges of the beam at a desirable location where the PI (hinge) must be located so as to secure the eventual ductile failure of the frame.

Figure 5.8 will probably provide the impetus to encourage the study of earthquakes, and their impact on man-made structures, to become a science instead of an art. It is clear that each horizontal vibration must create, in the given moment, the axial tension in the column and the alteration of the compression zone into the tensile zone of the connecting beam. As the result of the simultaneous action of the tensile forces T and the axial tension, or the transverse force V for the beam, the action of the punch force V_n will occur.

5.10.1 Concluding thoughts for the above discussion

If one accepts the proposition that the classical concept of diagonal tension is not correct, that all diagonal failures are due to the action of the internal active forces of compression, tension, and vertical or transverse shear; and that, in the course of an earthquake, upward-acting axial forces can appear in the column, it is not difficult to come to the following conclusion. By strengthening the connections of the column and the beam, catastrophic failure of frame structures by the corresponding internal active forces of compression, tension and shear through the column can be prevented for a given zone of the seismic area, thus enabling ductile failure to occur through the beam.

The connection of the beam to a column must satisfy the requirements of the laws of physics. These laws require that the beam, with its complete cross section, must penetrate the column in such a way that compression and tensile flanges are added between the column's flanges. These are known as "stiffeners" and enable the flexural compression and flexural tensile forces to act uninterrupted from one hinge (PI) to another. Further, the beam's web must be recompensated with two plates, located between the column's flanges and welded to the column's web and flanges. Thus the entire cross section of the beam is reestablished between the column's flanges in order to act as a continuous web from one beam's hinge (PI) to another.

If the above is not satisfied, then the largest flexural tensile and flexural compression forces (which are located precisely in flanges of the beam at the connecting point of the beam and the column) will cause stress concentration at the column's flanges which will lead to twisting of the flanges of the columns and beams, creating hinges at the connection of the beam and column. As a result, the so-called shear fracture will occur, as it happened during the Northridge, California earthquake in 1994.

Further, it is well known that the largest shear forces of the beam are located at the PIs, or hinges, but such hinges should not be closer to the column than the depth of the beam. This is a result of a stress distribution at 45°, where shear forces of a beam's load at the PIs (hinges) can influence the connecting surface of a beam and a column if they are closer than the depth of the beam. Yet, if they are closer, the sliding shear forces of a beam's load at the PI will add additional sliding stresses to the connecting surface at the column.

As a result of the fact that such a beam-to-column connection is very close to the location of the maximal negative moment of the beam, vertical shear forces at the same connection could be very close to zero. So, if it happens that such shear forces are very large, it must mean that there is a larger vertical load at the beam in the vicinity of this connection; or that something is fundamentally wrong concerning detailing of the connection. Also, we should bear in mind that the portion of the beam (between the hinges and the column's surfaces) should never be cantilevered (welded to the column's flanges without the so-called "stiffeners" between the column's flanges) because such a cantilever will cause full vertical shear forces precisely at the connecting surface of the beam and column. But if we have a corner of the building that requires a cantilever to exist, then besides the addition of the beam's flanges between the column's flanges (known as two stiffeners) and the beam's web, the entire column and beam's connection could be overlapped approximately 4 in from the connecting surface. These two plates should be able to take over the vertical shear forces at the connecting line of the cantilever and failure could never occur through the two connecting surfaces of the beam to the column, known as a "cantilever connection failure."

Special note should be made of the fact that the hinges (PI) must be secured and moved away from the column's face as far as rationally possible. Besides cutting flanges of the beam and making them smaller at the desired points where we want hinges (PI) to occur, the best solution, for now, is probably the application of a slotted beam to the column connection which "moves the plastic hinge region in the beam away from the face of the column."[29]

5.11 Design and prefabrication of shear wall panels for parallel testing using ACI 318-95 guidelines method and the triangular reinforcement method

5.11.1 Notation

$l_{w.tot}$	Length of entire wall or of segment of wall considered in direction of shear force including boundary elements.
l_b	Length of boundary element, web length.
b_b	Web width.
t_w	Thickness of flange.
f_{cp}	Specified compressive strength of concrete (psi).
f_y	Specified yield strength of reinforcement (psi).
h_w	Height of entire wall.
V_u	Applied shear force (kips) (k).
ϕ	Strength reduction factor.
M_u	Factored moment due to applied shear force.
P_u	Factored axial load at boundary element due to applied shear force.
I_g	Moment of inertia of web portion of wall.
A_{gb}	Area of boundary element.
$A_{st.b}$	Required area of steel (in^2).
A_{st}	Provided area of steel (in^2).
x	Rebar number.
A_{b_x}	Area of bar number x.
No_{bars}	Number of rebars.
x_h	Rebar number for the transverse reinforcement.
s	Required spacing of transverse reinforcement (in).

d_{b_m}	Diameter of transverse reinforcement.
h_c	Cross-sectional dimension of column core measured center-to-center of confining reinforcement.
f_{yh}	Yield strength of transverse reinforcement (psi).
t_{cover}	Provided concrete cover.
A_{ch}	Cross-sectional area of a structural member measured out-to-out of transverse reinforcement (in^2).
A_{sh}	Total cross-sectional area of transverse reinforcement within spacing, s, and perpendicular to dimension h_c.
$A_{sh.p}$	Provided cross-sectional area of transverse reinforcement.
A_{cv}	Net area of concrete section bounded by web thickness and length of section in the direction of shear force considered.
α_c	Coefficient defining the relative contribution of concrete strength to wall strength.
N_{wb}	Number of web rebars.
x_{wb}	Web rebar number.
s_{wb}	Spacing of web rebars.
A_{wb}	Provided area of web reinforcement.
V_n	Nominal shear (kips).
P_{dead}	Dead load (kips).
P_{live}	Live load (kips).
P_n	Maximum compressive load capacity (kips).
$P_o \cdot 0.8$	Column capacity if no slenderness or bending affecting the column (kips).
P_{Bal}	Balanced load (kips).
P_c	Critical buckling load (kips).
phi:	Factor per ACI 9.3.2.2.
C_m	Factor determined in accordance with ACI 10.11.5.3.
Ecc	Eccentricity.
K	Effective length factor for compression members.
L_u	Unsupported length.
r	Radius of gyration.

5.11.2 Samples for testing

Given: 5 in RC shear wall, 4 ft high and 5 ft long. Applied horizontal lateral load is 40 kips. Design the wall reinforcement and check the thickness using the following two methods (see Figures 5.9 and 5.10):

1   ACI 318-95 guidelines for reinforced concrete shear walls in seismic zones 3 & 4.
2   The newly proposed triangular reinforcement method.

5.11.2.1 Solution (1): ACI 318-95 method, section 21-6

Input data:

$$l_{w.tot} = 5\,\text{ft}, \qquad h_w = 4\,\text{ft}, \qquad t_w = 5\,\text{in}, \qquad V_u = 40\,\text{kips},$$
$$l_b = 6\,\text{in}, \qquad b_b = 5\,\text{in}, \qquad w_{conc} = 145\,\text{pcf},$$
$$f_{cp} = 5{,}000\,\text{psi}, \qquad f_y = 60{,}000\,\text{psi},$$
$$l_w = l_{w.tot} - l_b = 4.5\,\text{ft},$$
$$A_g = l_{w.tot} \cdot t_w = 2.083\,\text{ft}^2,$$
$$P_{dead} = t_w \cdot w_{conc} \cdot h_w \cdot l_{w.tot} = 1.208\,\text{kips}.$$

Check if boundary elements are required:

$$M_u = 1.4 \cdot V_u \cdot h_w,$$
$$P_{lu} = 1.4 \cdot P_{dead} = 1.692\,\text{kips},$$

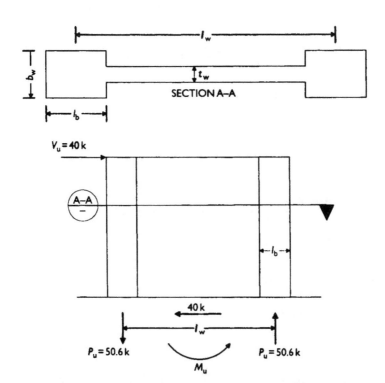

Figure 5.9 Wall elevation and section with boundary elements where needed.

$$I_g = \frac{t_w \cdot l_{w.tot}^3}{12} = 4.34 \, \text{ft}^4,$$

$$P_u = \frac{M_u}{l_w} + \frac{P_{lu}}{2} = 50.624 \, \text{kips}.$$

Calculate maximum compressive stress:

$$f_c = \frac{P_u}{A_g} + \frac{M_u \cdot (l_{w.tot}/2)}{I_g} = 1.065 \, \text{ksi}.$$

Check if boundary elements are required:

$$f_c = 1.065 \, \text{ksi} \gg 0.2 \cdot f_{cp} = 1 \, \text{ksi}.$$

Therefore, boundary elements are required.

Design of boundary elements:

$$l_b = 6 \, \text{in}, \qquad b_b = 5 \, \text{in}, \qquad \phi = 0.7,$$

$$A_{gb} = l_b \cdot b_b = 30 \, \text{in}^2.$$

Determine reinforcement by designing each boundary element as a short column:

$$A_{st.b} = \frac{(P_u/0.8 \cdot \phi) - 0.85 \cdot f_{cp} \cdot A_{gb}}{f_y - 0.85 \cdot f_{cp}} = -0.665 \, \text{in}^2,$$

$$\rho_b = \frac{A_{st.b}}{A_{gb}} = -0.022 \ll 0.06 \ (\text{OK}).$$

Try:

$$l_b = 6 \, \text{in}, \qquad b_b = 5 \, \text{in},$$

$$A_{gb} = l_b \cdot b_b = 30 \, \text{in}^2,$$

$$A_{st.b} = \frac{(P_u/0.8 \cdot \phi) - 0.85 \cdot f_{cp} \cdot A_{gb}}{f_y - 0.85 \cdot f_{cp}} = -0.665 \, \text{in}^2.$$

Enter rebar number $x = 3$, $No_{bars} = 4$.

Use 4–#3

$$A_{st} = A_{b_x} \cdot No_{bars} = 0.44 \, \text{in}^2,$$

$$\rho = \frac{A_{st}}{A_{gb}} = 1.467\%.$$

Check the required confinement reinforcement (maximum hoop spacing = 4 in):

$$\text{Try} \, \# \, x_h = 2 \, \text{hoops}, \qquad s = \text{if}\left(\frac{b_b}{4} \leq 4 \, \text{in}, \frac{b_b}{4}, 4 \, \text{in}\right) = 1.25 \, \text{in},$$

$$d_{b_{rh}} = 0.25 \, \text{in}, \qquad f_{yh} = 60,000 \, \text{psi}, \qquad t_{cover} = 0.5 \, \text{in},$$

$$h_c = l_b - 2 \cdot (t_{cover}) - d_{b_{r_h}} = 4.75\,\text{in},$$

$$A_{ch} = (h_c + d_{b_{r_h}}) \cdot (b_b - 2 \cdot t_{cover}) = 20\,\text{in}^2,$$

$$A_{sh} = \text{if}\left[0.09 \cdot s \cdot h_c \cdot \frac{f_{cp}}{f_{yh}} \geq 0.3 \cdot s \cdot h_c \cdot \left(\frac{A_{gb}}{A_{ch}} - 1\right) \cdot \frac{f_{cp}}{f_{yh}},\right.$$

$$\left. 0.09 \cdot s \cdot h_c \cdot \frac{f_{cp}}{f_{yh}}, \ 0.3 \cdot s \cdot h_c \cdot \left(\frac{A_{gb}}{A_{ch}} - 1\right) \cdot \frac{f_{cp}}{f_{yh}}\right]$$

$$= 0.074\,\text{in}^2.$$

Therefore, $\#x_h = 2$ hoops, with $N_t = 1$ crossties in each direction, provides:

$$A_{sh.p} = (2 + N_t) \cdot A_{b_{r_h}} = 0.15\,\text{in}^2 \gg A_{sh} = 0.074\,\text{in}^2 \ (\text{OK}).$$

Calculate the required web reinforcement, $V_u = 40\,\text{kips}$.
  Shear area:

$$A_{cv} = (l_w + l_b) \cdot t_w = 300\,\text{in}^2,$$

$$V_u = 40\,\text{kips} \ll 2 \cdot A_{cv} \cdot \sqrt{f_{cp}}\,(\text{psi}) = 42.426\,\text{kips}.$$

One curtain of reinforcement is required. Minimum longitudinal and transverse reinforcement ratio $= 0.0025$. Maximum bar spacing $= 18\,\text{in}$.
  Calculate the nominal shear strength:
  Calculate α_c:

$$\frac{h_w}{l_w + l_b} = 0.8 \leq 2,$$

$$\alpha_c = \text{if}\left[\frac{h_w}{l_w + l_b} \leq 1.5, \ 3.0, \text{if}\left[\frac{h_w}{l_w + l_b} \geq 2.0, 2.0, 6.0\right.\right.$$

$$\left.\left. - 2 \cdot \left(\frac{h_w}{l_w + l_b}\right)\right]\right] = 3,$$

$$A_{cv} = (l_w + l_b) \cdot t_w = 300\,\text{in}^2.$$

Assume $N_{wb} = 1\#$, $x_{wb} = 3$ at $s_{wb} = 10\,\text{in}$.

$$A_{rb} = A_{b_{rwb}} \cdot \frac{12\,\text{in}}{s_{wb}}, \qquad A_{wb} = A_{rb} \cdot N_{wb} = 0.132\,\text{in}^2,$$

$$\rho_{wb} = \frac{A_{wb}}{t_w \cdot 12\,\text{in}} = 2.2 \times 10^{-3},$$

$$\rho_n = \text{if}(\rho_{wb} \leq 0.0025, 0.0025, \rho_{wb}) = 2.5 \times 10^{-3}.$$

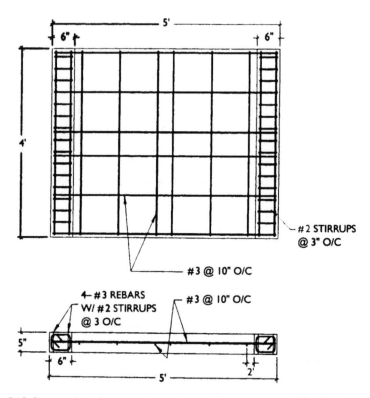

Figure 5.10 Shear wall reinforcement in section and elevation as per ACI 318-95 guidelines.

Calculate allowable shear load, $\phi = 0.7$.

$$\phi V_n = \phi \cdot A_{cv} \cdot \left[\alpha_c \cdot \sqrt{f_{cp} \ (\text{psi})} + \rho_n \cdot f_y \right]$$
$$= 76.048 \, \text{kips} \gg V_u = 40 \, \text{kips} \ (\text{OK}).$$

Expected failure of wall at $V_u := \phi V_n/\phi = 108.64 \, \text{kips}$.
Use $V_u = 108.64 \, \text{kips}$ for the design of wall using the triangular method.
Use minimum $\rho_v = \text{if}((h_w/l_w + l_b) \leq 2, \rho_n, 0.0025) = 2.5 \times 10^{-3}$.

$$A_{sv} = \rho_v \cdot t_w = 0.15 \, \text{in}^2/\text{ft}.$$

For $N_{wb} = 1\#$, $x_{wb} = 3$ at $s_{wb} = 10 \, \text{in}$,

$$S = \frac{A_{wb}}{A_{sw}} = 10.56 \, \text{in} \gg s_{wb} = 10 \, \text{in} \ (\text{OK}).$$

Use 1–#3 at 10 in.

5.11.2.2 Solution (2): triangular reinforcement method

The expected failure of the classical shear wall has been calculated to be at $V_u = 109$ kips (108.64 kips) and this force of 109 kips will be used to design the triangularly reinforced shear wall with the same dimensions – length 5 ft and height 4 ft.

For this method the wall is assumed to consist of members of truss elements, as shown. Each member is 5 in thick (the required thickness of the wall) and has a certain depth, as determined by the following calculation. A two-dimensional, finite-element frame analysis program (Table 5.1) was used to calculate loads on members of truss. From the analysis, two members were designed using an RC design program. Member 2–5 with a section of 5 in × 10 in is found to be adequate to resist an axial load of 85 kips (see Member End Forces of analysis) and member 2–3 with a section of 5 in × 6 in is found to be adequate to resist an axial load of 41.33 kips. See details for reinforcement and confinement (Figures 5.11–5.13 and Tables 5.2–5.4).

5.11.2.3 Sample for testing of classical shear wall under 109 kips

Given: 5 in RC shear wall, 4 ft high and 5 ft long. Applied horizontal lateral load is 109 kips. Design the wall reinforcement and check the thickness using the following two methods:

1 ACI 318-95 guidelines for reinforced concrete shear walls in seismic zones 3 & 4.
2 The newly proposed triangular reinforcement method.

5.11.2.4 Solution (3): ACI 318-95 method, section 21-6

Input data:

$$l_{w.tot} = 5\,\text{ft}, \qquad h_w = 4\,\text{ft}, \qquad t_w = 5\,\text{in}, \qquad V_u = 109\,\text{kips},$$

$$l_b = 9\,\text{in}, \qquad b_b = 5\,\text{in}, \qquad w_{conc} = 145\,\text{pcf},$$

$$f_{cp} = 5,000\,\text{psi}, \qquad f_y = 60,000\,\text{psi},$$

$$l_w = l_{w.tot} - l_b = 4.25\,\text{ft},$$

$$A_g = l_{w.tot} \cdot t_w = 2.083\,\text{ft}^2,$$

$$P_{dead} = t_w \cdot w_{conc} \cdot h_w \cdot l_{w.tot} = 1.208\,\text{kips}.$$

Check if boundary elements are required:

$$M_u = 1.4 \cdot V_u \cdot h_w,$$

$$P_{lu} = 1.4 \cdot P_{dead} = 1.692\,\text{kips},$$

$$I_g = \frac{t_w \cdot l_{w.tot}^3}{12} = 4.34\,\text{ft}^4,$$

$$P_u = \frac{M_u}{l_w} + \frac{P_{lu}}{2} = 144.469\,\text{kips}.$$

Table 5.1 Two-dimensional frame analysis

Nodes ...

Node label	Node coordinates		X restraint	Y restraint	Z restraint	Node temp (deg F)
	X (ft)	Y (ft)				
1	0.000	0.000	Fixed	Fixed		0
2	0.000	4.000				0
3	5.000	4.000				0
4	5.000	0.000	Fixed	Fixed		0
5	2.500	2.000				0

Member ...

Member label	Property label	Endpoint nodes		Member length (ft)	I end releases			J end releases		
		I node	J node		X	Y	Z	X	Y	Z
1–2	Column	1	2	4.000						
1–5	Brace	1	5	3.202			Free			
2–3	Beam	2	3	5.000			Free			Free
2–5	Brace	2	5	3.202			Free			
3–4	Column	3	4	4.000						
3–5	Brace	3	5	3.202			Free			
4–5	Brace	4	5	3.202			Free			

Materials ...

Member label	Youngs (ksi)	Density (kcf)	Thermal (in/100d)	Yield (ksi)
Concrete	5,098.00	0.150	0.000000	60.00
Default	1.00	0.000	0.000000	1.00
Steel	29,000.00	0.490	0.000650	36.00

Section sections ...

Prop label Group tag	Material	Area	Depth width	Tf Tw	Ixx Iyy
Column	Concrete	30.000 in²	6.000 in	0.000 in	90.00 in⁴
			5.000 in	0.000 in	0.00 in⁴
Beam	Concrete	30.000 in²	6.000 in	0.000 in	90.00 in⁴
			5.000 in	0.000 in	0.00 in⁴
Brace	Concrete	50.000 in²	10.000 in	0.000 in	416.67 in⁴
			5.000 in	0.000 in	1.00 in⁴
Default	Default	1.000 in²	0.000 in	0.000 in	1.00 in⁴
			0.000 in	0.000 in	0.00 in⁴

Table 5.1 (Continued)

Node loads ...

Node label	Concentrated loads and moments			Load case factors				
	X	Y	Moment	#1	#2	#3	#4	#5
2	109.000 k			1.000				

Load combinations ...

Load combination description	Stress increase	Gravity load factors		Load combination factors				
		X	Y	#1	#2	#3	#4	#5
Basic	1.000			1.000				

Node displacements & reactions

Node label	Load combi- nation	Node displacements			Node reactions		
		X (in)	Y (in)	Z (Radians)	X (k)	Y (k)	Z (k ft)
1	Basic	0	0	−0.00095	−43.60446	−87.20000	0
2	Basic	0.04569	0.01616	−0.00095	0	0	0
3	Basic	0.02948	−0.01045	−0.00061	0	0	0
4	Basic	0	0	−0.00061	−65.39554	87.20000	0
5	Basic	0.01362	−0.00342	−0.00044	0	0	0

Member end forces ...

Member label	Load combi- nation	Node "I" end forces			Node "J" end forces		
		Axial (k)	Shear (k)	Moment (ft k)	Axial (k)	Shear (k)	Moment (ft k)
1–2	Basic	−51.48852	0	0	51.48852	0	0
1–5	Basic	−56.35815	−0.64649	0	56.35815	0.64649	−2.06977
2–3	Basic	41.33005	0	0	−41.33005	0	0
2–5	Basic	85.00598	2.06730	0	−85.00598	−2.06730	6.61860
3–4	Basic	33.28598	0	0	−33.28598	0	0
3–5	Basic	−53.06693	−0.17331	0	53.06693	0.17331	−0.55486
4–5	Basic	84.74516	−1.24751	0	−84.74516	1.24751	−3.99397

Table 5.1 (Continued)

Member overall envelope summary

Member label	Section	Axial (k)	Shear (k)	Moment (ft k)	Deflection (in)
1–2	Column	51.489			0.046
1–5	Brace	56.358	0.646	2.070	0.011
2–3	Beam	41.330			0.026
2–5	Brace	85.006	2.067	6.619	0.035
3–4	Column	33.286			0.029
3–5	Brace	53.067	0.173	0.555	0.015
4–5	Brace	84.745	1.248	3.994	0.006

Deflection values listed are the maximum of a sampling of 31 points across the member.

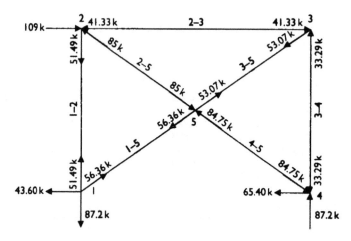

Figure 5.11 Frame members and axial loads from the analysis using the triangular reinforcement method.

Calculate maximum compressive stress:

$$f_c = \frac{P_u}{A_g} + \frac{M_u \cdot (l_{w.tot}/2)}{I_g} = 2.923 \text{ ksi.}$$

Check if boundary elements are required:

$$f_c = 2.923 \text{ ksi} \gg 0.2 \cdot f_{cp} = 1 \text{ ksi.}$$

Therefore, boundary elements are required.
  Design of boundary elements:

$$l_b = 9 \text{ in}, \qquad b_b = 5 \text{ in}, \qquad \phi = 0.7.$$
$$A_{gb} = l_b \cdot b_b = 45 \text{ in}^2.$$

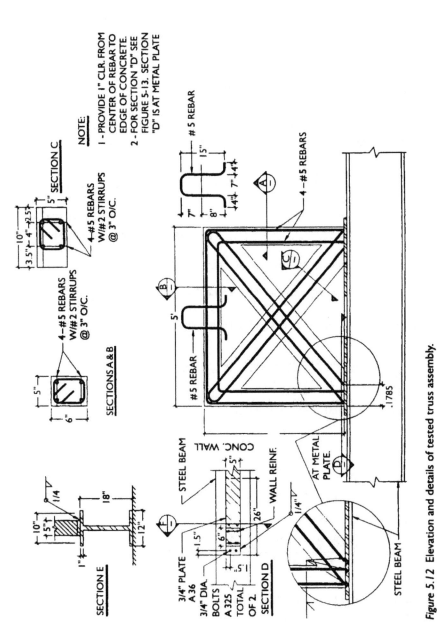

Figure 5.12 Elevation and details of tested truss assembly.

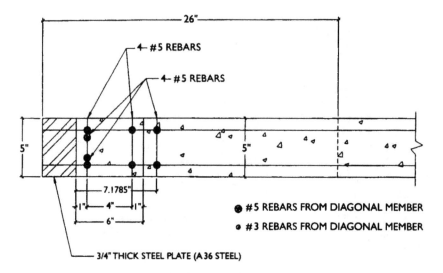

Figure 5.13 Specific design details for Figure 5.12, Section D.

Determine reinforcement by designing each boundary element as a short column:

$$A_{st.b} = \frac{(P_u/0.8 \cdot \phi) - 0.85 \cdot f_{cp} \cdot A_{gb}}{f_y - 0.85 \cdot f_{cp}} = 1.197 \, in^2.$$

$$\rho_b = \frac{A_{st.b}}{A_{gb}} = 0.027 \ll 0.06 \ (OK).$$

Try:

$$l_b = 9 \, in, \qquad b_b = 5 \, in,$$

$$A_{gb} = l_b \cdot b_b = 45 \, in^2,$$

$$A_{st.b} = \frac{(P_u/0.8 \cdot \phi) - 0.85 \cdot f_{cp} \cdot A_{gb}}{f_y - 0.85 \cdot f_{cp}} = 1.197 \, in^2.$$

Enter rebar number $x = 5$, No$_{bars} = 4$.
Use 4–#5

$$A_{st} = A_{b_x} \cdot No_{bars} = 1.24 \, in^2.$$

$$\rho = \frac{A_{st}}{A_{gb}} = 2.756\%.$$

Check the required confinement reinforcement (maximum hoop spacing = 4 in):

Try $\# x_h = 2$ hoops, $s = \text{if}\left(\dfrac{b_b}{4} \le 4\,\text{in}, \dfrac{b_b}{4}, 4\,\text{in}\right) = 1.25\,\text{in},$

$d_{b_h} = 0.25\,\text{in},$ $f_{yh} = 60{,}000\,\text{psi},$ $t_{cover} = 0.5\,\text{in},$

$h_c = l_b - 2 \cdot (t_{cover}) - d_{b_h} = 7.75\,\text{in},$

$A_{ch} = (h_c + d_{b_h}) \cdot (b_b - 2 \cdot t_{cover}) = 32\,\text{in}^2,$

$A_{sh} = \text{if}\left[0.09 \cdot s \cdot h_c \cdot \dfrac{f_{cp}}{f_{yh}} \ge 0.3 \cdot s \cdot h_c \cdot \left(\dfrac{A_{gb}}{A_{ch}} - 1\right) \cdot \dfrac{f_{cp}}{f_{yh}},\right.$

$\left. 0.09 \cdot s \cdot h_c \cdot \dfrac{f_{cp}}{f_{yh}}, \; 0.3 \cdot s \cdot h_c \cdot \left(\dfrac{A_{gb}}{A_{ch}} - 1\right) \cdot \dfrac{f_{cp}}{f_{yh}}\right]$

$= 0.098\,\text{in}^2.$

Therefore, $\# x_h = 2$ hoops, with $N_t = 1$ crossties in each direction, provides:

$A_{sh.p} = (2 + N_t) \cdot A_{b_h} = 0.15\,\text{in}^2 \gg A_{sh} = 0.098\,\text{in}^2$ (OK).

Calculate required web reinforcement, $V_u = 109\,\text{kips}.$
Shear area:

$A_{cv} = (l_w + l_b) \cdot t_w = 300\,\text{in}^2,$

$V_u = 109\,\text{kips} \gg 2 \cdot A_{cv} \cdot \sqrt{f_{cp}}\,\text{(psi)} = 42.426\,\text{kips}.$

Two curtains of reinforcement are required. Minimum longitudinal and transverse reinforcement ratio = 0.0025. Maximum bar spacing = 18 in.

Calculate nominal shear strength:
Calculate α_c:

$\dfrac{h_w}{l_w + l_b} = 0.8 \le 2,$

$\alpha_c = \text{if}\left[\dfrac{h_w}{l_w + l_b} \le 1.5, 3.0, \text{if}\left[\dfrac{h_w}{l_w + l_b} \ge 2.0, 2.0, 6.0\right.\right.$

$\left.\left. -2 \cdot \left(\dfrac{h_w}{l_w + l_b}\right)\right]\right] = 3,$

$A_{cv} = (l_w + l_b) \cdot t_w = 300\,\text{in}^2.$

Table 5.2 Rectangular concrete column – description, member 1–2

General information

Width	5.000 in	fc	5,000.0 psi	Unbraced length	4.000 ft
Depth	6.000 in	Fy	60,000.0 psi	Eff. length factor	1.000
Rebar:		Seismic zone	4	Column is UNBRACED	
2–#5d = 1.000 in		LL & ST loads act		Delta:S	1.00
2–#5d = 5.000 in		separately			

Loads

Axial loads	Dead Load (k)	Live Load (k)	Short Term 51.490 k	Eccentricity (in)
Summary				Column is OK

5.00 × 6.00 in column, rebar: 2–#5 @ 1.00 in, 2–#5 @ 5.00 in

	ACI 9-1	ACI 9-2	ACI 9-3	
Applied: P_u: Max factored	0.00 k	72.09 k	72.09 k	
Allowable: P_n Phi @ Design Ecc	101.82 k	98.20 k	98.20 k	
M-critical	0.00 k ft	4.69 k ft	4.69 k ft	
Combined eccentricity	0.780 in	0.780 in	0.780 in	
Magnification factor	1.00	1.12	1.12	
Design eccentricity	0.780 in	0.877 in	0.877 in	
Magnified design moment	0.00 k ft	5.27 k ft	5.27 k ft	
P_o * .80	157.30 k	157.30 k	157.30 k	
P: Balanced	43.38 k	43.38 k	43.38 k	
Ecc: Balanced	5.230 in	5.230 in	5.230 in	

Slenderness

Actual k L_u/r 26.667	Elastic Modulus	4,030.5 ksi	Beta	0.800
	ACI Eq. 9-1	ACI Eq. 9-2	ACI Eq. 9-3	
Neutral axis distance	6.1039 in	5.8929 in	5.8929 in	
Phi	0.7000	0.7000	0.7000	
Max limit kl/r	22.0000	22.0000	22.0000	
Beta = M:sustained/M:max	0.0000	0.0000	0.0000	
C_m	1.0000	1.0000	1.0000	
EI/1000	216.39	216.39	216.39	
P_c : pi^2 E I/(k L_u)^2	926.94	926.94	926.94	
alpha: MaxP$_u$/(phi P$_c$)	0.0000	0.1111	0.1111	
Delta	1.0000	1.1250	1.1250	
Ecc: Ecc loads + moments	0.7800	0.7800	0.7800 in	
Design Ecc = Ecc Delta	0.7800	0.8775	0.8775 in	

ACI factors (per ACI, applied internally to entered loads)

ACI 9-1 & 9-2 DL	1.400	ACI 9-2 group factor	0.750	UBC 1921.2.7 "1.4" factor	1.400
ACI 9-1 & 9-2 LL	1.700	ACI 9-3 dead load factor	0.900	UBC 1921.2.7 "0.9" factor	0.900
ACI 9-1 & 9-2 ST	1.700	ACI 9-3 short term factor	1.300		
... seismic = ST :	1.100				

Table 5.3 Rectangular concrete column – description, member 2–3

General information

Width	5.000 in	fc	5,000.0 psi	Unbraced length	5.000 ft
Depth	6.000 in	Fy	60,000.0 psi	Eff. length factor	1.000
Rebar:		Seismic zone	4	Column is UNBRACED	
2–#5 d = 1.000 in		LL & ST loads act separately		Delta:S	1.00
2–#5 d = 5.000 in					

Loads

Axial loads	Dead load (k)	Live load (k)	Short term 41.330 k	Eccentricity (in)
				Column is OK

Summary
5.00 × 6.00 in column, rebar: 2–#5 @ 1.00 in, 2–#5 @ 5.00 in

	ACI 9-1	ACI 9-2	ACI 9-3	
Applied: P$_u$: Max factored	0.00 k	57.86 k	57.86 k	
Allowable: P$_n$ * Phi @ Design Ecc.	101.82 k	97.15 k	97.15 k	
M-critical	0.00 kft	3.76 kft	3.76 kft	
Combined eccentricity	0.780 in	0.780 in	0.780 in	
Magnification factor	1.00	1.16	1.16	
Design eccentricity	0.780 in	0.906 in	0.906 in	
Magnified design moment	0.00 kft	4.37 kft	4.37 kft	
P$_o$ * .80	157.30 k	157.30 k	157.30 k	
P: Balanced	43.38 k	43.38 k	43.38 k	
Ecc: Balanced	5.230 in	5.230 in	5.230 in	

Slenderness

Actual k L_u/r 33.333	Elastic modulus	4,030.5 ksi	Beta	0.800

	ACI Eq. 9-1	ACI Eq. 9-2	ACI Eq. 9-3
Neutral axis distance	6.1039 in	5.8326 in	5.8326 in
Phi	0.7000	0.7000	0.7000
Max Limit kl/r	22.0000	22.0000	22.0000
Beta = M:sustained/M:max	0.0000	0.0000	0.0000
C_m	1.0000	1.0000	1.0000
EI/1000	216.39	216.39	216.39
P_c: pi^2 E I/(k L_u)^2	593.24	593.24	593.24
alpha: MaxP_u/(phi P_c)	0.0000	0.1393	0.1393
Delta	1.0000	1.1619	1.1619
Ecc: Ecc loads + moments	0.7800	0.7800 in	0.7800 in
Design Ecc = Ecc * Delta	0.7800	0.9063	0.9063 in

ACI factors (per ACI, applied internally to entered loads)

ACI 9-1 & 9-2 DL	1.400	ACI 9-2 group factor	0.750	UBC 1921.2.7 "1.4" factor	1.400
ACI 9-1 & 9-2 LL	1.700	ACI 9-3 dead load factor	0.900	UBC 1921.2.7 "0.9" factor	0.900
ACI 9-1 & 9-2 ST	1.700	ACI 9-3 short term factor	1.300		
...seismic = ST':	1.100				

Table 5.4 Rectangular concrete column – description, member 2–5

General information

Width	5.000 in	fc	5,000.0 psi	Unbraced length	6.400 ft
Depth	10.000 in	Fy	60,000.0 psi	Eff. length factor	1.000
Rebar:		Seismic zone	4	Column is UNBRACED	
2-#5 d = 2.500 in		LL & ST loads act separately		Delta:S	1.00
2-#5 d = 6.500 in					
Loads					

Axial loads	Dead load (k)	Live load (k)	Short term (k)	Eccentricity (in)
Applied moments				
@ Top	5.000 kft	kft	kft	
@ Bottom	kft	kft	kft	

Summary   Column is OK
5.00 × 10.00 in column, rebar: 2-#5 @ 2.50 in, 2-#5 @ 6.50 in

	ACI 9-1	ACI 9-2	ACI 9-3
Applied: P_u: Max factored	0.00 k	119.00 k	119.00 k
Allowable: P_n * Phi @ Design Ecc	157.71 k	124.08 k	150.07 k
M-critical	0.00 kft	15.25 kft	9.80 kft
Combined eccentricity	0.900 in	1.538 in	0.988 in
Magnification factor	1.00	1.18	1.18
Design eccentricity	0.900 in	1.812 in	1.165 in
Magnified design moment	0.00 kft	17.96 kft	11.55 kft
P_o * .80	225.30 k	225.30 k	225.30 k
P: Balanced	40.76 k	40.76 k	40.76 k
Ecc: Balanced	7.603 in	7.603 in	7.603 in

Slenderness

Actual k L_u/r 25.600	Elastic modulus	4,030.5 ksi	Beta	0.800
	ACI Eq. 9-1	ACI Eq. 9-2	ACI Eq. 9-3	
Neutral axis distance	10.4406 in	7.9727 in	9.6888 in	
Phi	0.7000	0.7000	0.7000	
Max limit kl/r	22.0000	22.0000	22.0000	
Beta = M:sustained/M:max	0.0000	0.0000	0.0000	
C_m	1.0000	1.0000	1.0000	
EI/1000	671.75	671.75	671.75	
P_c: pi^2 E I/(k L_u)^2	1,124.05	1,124.05	1,124.05	
alpha: MaxP_u/(phi P_c)	0.0000	0.1512	0.1512	
Delta	1.0000	1.1782	0.1782	
Ecc: Ecc loads + moments	0.9000	1.5375	0.9884 in	
Design Ecc = Ecc * Delta	0.9000	1.8115	1.1645 in	

ACI factors (per ACI, applied internally to entered loads)

ACI 9-1 & 9-2 DL	1.400	ACI 9-2 group factor	0.750	UBC 1921.2.7 "1.4" factor	1.400
ACI 9-1 & 9-2 LL	1.700	ACI 9-3 dead load factor	0.900	UBC 1921.2.7 "0.9" factor	0.900
ACI 9-1 & 9-2 ST	1.700	ACI 9-3 short term factor	1.300		
...seismic = ST'	1.100				

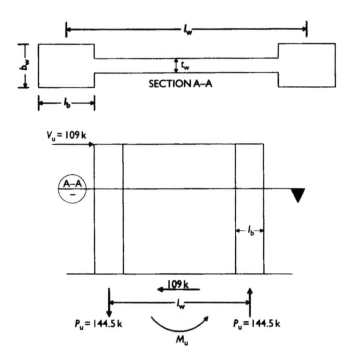

Figure 5.14 Wall elevation and section with boundary elements where needed.

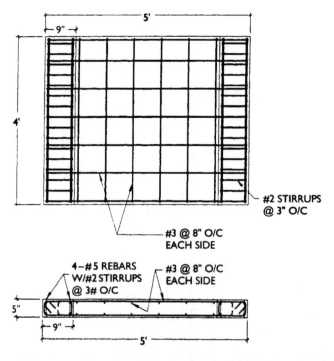

Figure 5.15 Shear wall reinforcement in section and elevation as per ACI 318-95 guidelines.

Assume $N_{wb} = 2\#$, $x_{wb} = 3$ at $s_{wb} = 8$ in.

$$A_{rb} = A_{b_{z_{wb}}} \cdot \frac{12 \text{ in}}{s_{wb}}, \qquad A_{wb} = A_{rb} \cdot N_{wb} = A_{wb} = 0.33 \text{ in}^2,$$

$$\rho_{wb} = \frac{A_{wb}}{t_w \cdot 12 \text{ in}} = 5.5 \times 10^{-3},$$

$$\rho_n = \text{if } (\rho_{wb} \le 0.0025, 0.0025, \rho_{wb}) = 5.5 \times 10^{-3}.$$

Calculate allowable shear load, $\phi = 0.7$.

$$\phi V_n = \phi \cdot A_{cv} \cdot \left[\alpha_c \cdot \sqrt{f_{cp} \text{ (psi)}} + \rho_n \cdot f_y \right]$$

$$= 113.848 \text{ kips} \gg V_u = 109 \text{ kips (OK)}.$$

Expected failure of wall at: $V_u := \phi V_n / \phi$, $V_u = 162.64$ kips.
Use $V_u = 162.64$ kips for the design of wall using the triangular method.
Use minimum $\rho_v = \text{if } (h_w/(l_w + l_b) \le 2, \rho_n, 0.0025) = 5.5 \times 10^{-3}$.

$$A_{sv} = \rho_v \cdot t_w = 0.33 \text{ in}^2/\text{ft}.$$

For $N_{wb} = 2\#$, $x_{wb} = 3$ at $s_{wb} = 8$ in,

$$S = \frac{A_{wb}}{A_{sv}} = 12 \text{ in} \gg s_{wb} = 8 \text{ in (OK)}.$$

Use 2–#3 at 8 in.

5.11.3 Experimental testing of old and new shear walls

5.11.3.1 General observations

In order to find a correlation between the behavior of the existing shear wall system and the new system, a classical shear wall was designed for horizontal shear force $V_u = 40$ kips. For the triangularly reinforced shear wall, a force will be used which will cause ultimate failure determined by structural analyses when the classical shear wall is exposed to 40 kips of loading. In this case, it has been determined that this force is $V_u = 108.84$ kips or 109 kips approximately.

For this reason, a horizontal force of 109 kips will be used to design a new triangularly reinforced shear wall and comparison will be made between the economical and the engineering safety points of view for the existing and new systems. Eventually, if the new system becomes relatively less expensive and much safer, the classical concept of shear walls should be abandoned in favor of the new shear wall throughout the world. The motive of the next test will be to find out what correlation exists from the economical and safety points of view if one designs a third panel as per the classical method and exposes it to the same horizontal shear force of 109 kips exactly as the new shear wall has been designed. The same comparative analysis will be applied as has been done for the first case, to find out how these two systems fare on the economical and safety fronts.

5.11.3.2   Old and new shear walls exposed to different loads

A classical panel with a horizontal load of 40 kips versus a triangularly reinforced panel with a horizontal load of 109 kips (as per design in Section 5.11.2.1 versus that in Section 5.11.2.2) have been investigated. The classical panel, marked C-2, 5 ft long, 4 ft high, 5 in thick, 5,000 psi concrete, total reinforcement of 43.6 lb, has been designed for 40 kips horizontal load and failure occurs at a force of 67.3 kips. The total absorbed load per pound of reinforcement is 67,300/43.6 = 1,543 lb/#.

The triangularly reinforced panel, marked X-1 with the same dimensions as the classical panel with total reinforcement of 125 lb, has been designed for 109 kips horizontal load while its failure occurs at 220 kips. The total absorption of load per pound of reinforcement is 220,000/125 = 1,760 lb/#.

5.11.3.2.1   ECONOMICAL COMPARISON

Panel C-2 absorbed 1,543 lb per pound of its reinforcement while panel X-1 absorbed 1,760 lb/#. This means that the new system is more economical by 14% even for this load.

5.11.3.2.2   SAFETY COMPARISON

Panel C-2 failed at 67.3 kips while panel X-1 failed at 220 kips, which means that the new system is safer by 326.9 percent, or 3.3 times safer than the old system.

5.11.3.3   Old and new shear walls exposed to identical loads

The classical panel with a horizontal load of 109 kips versus the triangularly reinforced panel with a load of 109 kips have been investigated and are as described in Sections 5.11.2.3 and 5.11.2.2 of this chapter.

The classical panel, marked C-1, with the same dimensions, 5 ft long, 4 ft high, 5 in thick, 5,000 psi concrete, total reinforcement of 94 lb, has been designed for a horizontal load of 109 kips and failure occurs at 110 kips. The total absorbed load per pound of reinforcement is 110,000/94 = 1,170 lb/#.

The triangularly reinforced panel, marked X-1, with the same dimensions as the classical panel, and with total reinforcement of 125 lb, has been designed for 109 kips horizontal load while its failure occurs at 220 kips of external loading. The total absorption of load per pound of reinforcement is 222,000/125 = 1,760 lb/#.

5.11.3.3.1 ECONOMICAL COMPARISON

Panel C-1 absorbed 1,170 lb/# of its reinforcement while panel X-1 absorbed 1,760 lb/#. That means that the new system is more economical than the classical system by $1,760 \times 100/1,170 = 150\%$ or 1.5 times better.

5.11.3.3.2 SAFETY COMPARISON

Panel C-1 failed at 110 kips while panel X-1 failed at 220 kips, which means that the new system is safer by $220,000 \times 100/110,000 = 200\%$ or two times safer than the old systems.

5.11.3.4 Remarks

The new system is recommended not only because it is economically advantageous, but also because it achieves higher safety of the shear wall with its inherent ductility of failure. Yet, it should not be the final word on the new system versus the old system; rather it should remain the task for future researchers to find out additional information in practical applications of the new system, which is beyond any shadow of doubt superior to the classical concept. The main reason for its superiority is that the new concept is based on a natural law of physics that diagonal forces can be controlled by diagonal reinforcement, while the classical concept is based on the diagonal tension theory, which is incorrect (as is proven in this book).

The evolution of rational thought is a characteristic that elevates mankind and allows for advances in all areas of knowledge. Just as we developed the wheel as an advancement over simply dragging objects, we can step forward and use this new system of diagonal reinforcement as an advancement over the classical concept of reinforcement when designing shear walls to resist seismic forces. Consequently, there is no reasonable justification for any comparison between these two systems, because one is simple illusion while the other is based on the laws of physics. Rather, we should modify and improve applications of the new system in different ways as, for example, suggested in Section 5.11.6.3.

5.11.4 Replacement of higher-grade steel with lower-grade steel for the same panel

Since it is the first time in the history of engineering science that a triangularly RC shear wall, with its inherent ductility, has been tested, we believe that it will be very informative, for the same panel and the same design of 109 kips, to compare ultimate loads with reinforcing simply changed from 60 grade to 40 grade.

The reason for this is that the lower grade possesses higher plasticity with higher ductility; and it will be informative to compare the ultimate strength of the same panel reinforced by the higher grade versus the same quantity of lower grade steel.

We will now compare the classical panel with the higher grade of steel with its ultimate loads, and the new design with a lower-grade steel and its ultimate loads.

Unfortunately, panel X-2 did not fail in diagonal compression nor in diagonal tension, but rather, it was incapacitated for further testing by punching shear failure at the loading corner of the panel by the compressive element of the hydraulic ram which was only 4.5 in in diameter. We needed a minimum of 5 in $\times$ 10 in $= 50\,\text{in}^2$ for possible failure at 250 kips.

Yet, failure at punch shear occurred at 200 kips. Unfortunately, this detail of a minimum required loading area on the concrete panel of 5,000 psi was overlooked by the laboratory experts of the Smith Emery Company, as well as by us – we did not check this detail before loading the panel.

By comparing panels X-1 and X-2, it follows that if panel X-1, with grade 40 reinforcement, failed at 220 kips, panel X-2, with the same quantity of rein- forcement and grade 60, should fail at $220 \times 1.33 \cong 290$ kips. But even this very poor result of 200 kips is an excellent support for the new idea of triangular reinforcement of any shear wall.

Evidently, the idea of comparing these two panels (X-1 and X-2) with the same quantities but different quality of reinforcement, remains to be explored.

5.11.5 Official report from testing company

The testing for all panels was conducted by the Smith Emery Laboratories of Los Angeles, California. The details of their report are illustrated below. (See Figures 5.16–5.26.)

5.11.6 Design examples of multistory shear wall using the ACI 318-95 guidelines method and the triangular reinforcement method

Given: Five-story building with 6 in RC shear wall, 50 ft high and 32 ft long. Floors are 10 ft apart. Applied horizontal loads are assumed to be combined lateral and torsional loads. See Figure 5.28 for loads applied. For simplicity, live load is assumed to be zero and the only acting dead load is the wall self weight. Design the wall reinforcement and check the thickness using the following two methods:

1   ACI 318-95 guidelines for reinforced concrete shear walls in seismic zones 3 & 4.
2   The newly proposed triangular reinforcement method.

Design of horizontal beams and confinement reinforcement is beyond the scope of this example. This example is presented to illustrate the validity of the triangular reinforcement method.

SMITH-EMERY LABORATORIES
An Independent Commercial Testing Laboratory, Established 1904

781 E. Washington Boulevard
Los Angeles, California 90021
(213) 749-3411
Fax (213) 741-8626

File No   34687   September 21, 1999
Lab No   L-99-933

Client   HRISTA STAMENKOVIC
          Riverside, CA 92501-3415

Subject   **Shear Tests on 5,000 PSI Concrete Panel (4'x5'x5" thick) Identified as C-1 & C-2**
          (Conventional Reinforcement ACI 318-95) and X-1 & X-2 (Triangular Reinforcement).
Specifications: Load Test per ASTM E 72 Test Method
Source   Submitted to Laboratory by Client.

Report of Tests

SCOPE: To conduct a shear test on your concrete panel. The purpose of our testing was to evaluate the
         static shear capacity of a typical ACI 318-95 reinforcement compared to "X" experimental
         triangular system, and to provide data for the determination of the stiffness of the construction

FABRICATION: *(Please see attached Clients' detail drawings and photos)*
         Rebar reinforcements (all Grade 60, except X-1 which is Grade 40) were welded on a 5" wide by
         5'-6" long and 3/4" thick steel plate that serves as the base; concrete then was poured to form the
         4' high by 5' wide and 5" thick panel. One system is following the ACI 318-95 reinforcement
         guideline while another system is experimental "Triangular" reinforcement. Concrete strength was
         assumed at 5,000 PSI as per Client.

PROCEDURE:
         Shear test were conducted using a calibrated hydraulic ram system with manually controlled
         electric hydraulic pump, an electronic pressure transducer gage by Omega Engineering Model
         PX931-10KSV (with Enerpac 0-10,000 PSI analog pressure gage back-up), Hewlett Packard
         Data Acquisition Unit Model # 3497 A, IBM format computer for data processing, and 4 pieces
         of Differential Resistance Displacement Transducer by Celesco at 5" maximum stroke (with a
         resolution of 0.001"). Displacement are monitored at 0.001" resolution

         The panel were subjected to three (3) sections of "Horizontal Lateral" loading and unloading; first
         at 1/3 of expected ultimate load, second at 2/3, and the third at final loading until failure occurs
         Load (applied on a 6"x14"x 1.5" steel plate) and deflection are recorded at all phase of the testing

RESULTS:
         By applying the above procedure, ultimate failure for individual panels have been as follows
         1 Panel C-1 failed at maximum load of 110 KIPS
         2. Panel C-2 failed at maximum load of 67.3 KIPS
         3. Panel X-1 failed at maximum load of 220 KIPS
         4. Panel X-2 was suspended for further testing at 200 KIPS

Respectfully Submitted,   *(Please see attached graph and data)*
SMITH-EMERY COMPANY

JAMES E PARTRIDGE
President
Registered Civil Engineer No. 25270
Registration Expires 12-31-0
JEP:rc

25270

SMITH-EMERY LABORATORIES
An Independent Commercial Testing Laboratory, Established 1904

781 E. Washington Boulevard
Los Angeles, California 90021   *(213) 749-3411 Fax (213) 741-8626*

File No.:   34687
Lab No   L-99-933

Summarized Test Results

(-) Compression
(+) Tension

A. Panel C-1   Maximum Design Load, KIPS = 109 KIPS
                        Maximum Actual Load, KIPS = 110 KIPS

Load, lbs.	Deflection, inches			
	TR #1	*TR #2*	*TR #3*	*TR #4*
30,700	0.000	0.000	-0.140	0.014
75,400	0.000	0.000	-0.300	0.031
100,000	0.020	-0.010	-0.430	0.081
110,000	0.030	-0.010	-0.510	0.141	Maximum Load

B. Panel C-2   Maximum Design Load, KIPS = 40 KIPS
                        Maximum Actual Load, KIPS = 67.3 KIPS

Load, lbs.	Deflection, inches			
	TR #1	*TR #2*	*TR #3*	*TR #4*
15,500	0.000	0.000	-0.060	0.016
30,600	0.000	0.000	-0.200	0.104
40,100	0.010	0.000	-0.340	0.187
67,300	0.340	-0.010	-0.910	0.274	Maximum Load

C. Panel X-1   Maximum Design Load, KIPS = 109 KIPS
                        Maximum Actual Load, KIPS = 220 KIPS

Load, lbs.	Deflection, inches			
	TR #1	*TR #2*	*TR #3*	*TR #4*
35,100	0.000	0.000	0.010	-0.160
72,100	0.000	0.000	0.020	-0.270
140,000	-0.010	0.040	0.070	-0.480
176,000	-0.010	0.140	0.200	-0.860	4th Loading

D. Panel X-2   Maximum Design Load, KIPS = 109 KIPS
                        Maximum Actual Load, KIPS = More than 200 KIPS

Load, lbs.	Deflection, inches			
	TR #1	*TR #2*	*TR #3*	*TR #4*
30,900	-0.005	0.080	-0.010	0.000
56,354	-0.015	0.162	-0.008	0.011
99,352	-0.029	0.302	-0.049	0.016
146,992	-0.082	0.505	-	0.051	4th Loading

Figure 5.16 Classical reinforcement for panel C-2.

Figure 5.17 Classical reinforcement for panel C-2 in position to be concreted.

Figure 5.18 Triangular reinforcement for panel X-2 (X-1 panel is identical to X-2, except that the grade of steel is now 40,000 psi while for X-2 is 60,000 psi).

Figure 5.19 Triangular reinforcement for panel X-2 (same for X-1) in position to be concreted.

Figure 5.20 Classical reinforcement for panel C-1 in position to be concreted.

Figure 5.21 Panel C-2 at the end of testing.

Figure 5.22 Panel C-1 during testing procedure.

Figure 5.23 Panel X-2 during testing procedure.

Figure 5.24 Panel X-1 during testing procedure.

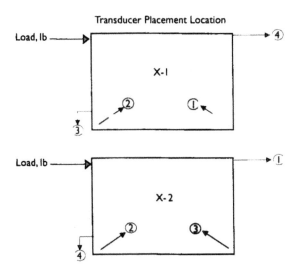

Figure 5.25 Location of tensometers during testing procedure (panels X-1, X-2).

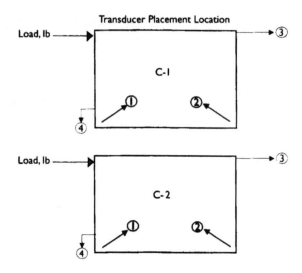

Figure 5.26 Location of tensometers during testing procedure (panels C-1, C-2).

5.11.6.1 Solution (1): ACI 318-95 method, section 21-6

Input data:

$$l_{w.tot} = 32\,ft, \qquad h_w = 50\,ft, \qquad t_w = 6\,in,$$
$$V = 91\,kips, \qquad M = 3{,}278\,kips\,ft,$$
$$V_u = 1.4\,V, \qquad M_u = 1.4\,M,$$
$$l_b = 26\,in, \qquad b_b = 6\,in, \qquad w_{conc} = 145\,pcf,$$
$$P_{dead} = t_w \cdot w_{conc} \cdot h_w \cdot l_{w.tot} = 116\,kips, \qquad P_{live} = 0\,kips,$$
$$f_{cp} = 10{,}000\,psi, \qquad f_y = 60{,}000\,psi,$$
$$l_w = l_{w.tot} - l_b = 29.833\,ft,$$
$$A_g = l_{w.tot} \cdot t_w = 16\,ft^2.$$

Check if boundary elements are required:

$$M_u = 4.589 \times 10^3 \,kips\,ft,$$
$$P_{lu} = 1.4 P_{dead} + 1.7 P_{live} = 162.4\,kips,$$

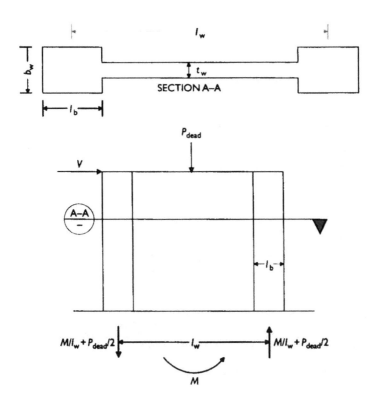

Figure 5.27 Wall elevation, section, boundary elements where needed and applied service loads.

$$I_g = \frac{t_w \cdot l_{w.tot}^3}{12} = 1.365 \times 10^3 \text{ ft}^4,$$

$$P_u = \frac{M_u}{l_w} \cdot 1.4 + \frac{P_{lu}}{2} = 296.559 \text{ kips}.$$

Calculate the maximum compressive stress:

$$f_c = \frac{P_u}{A_g} + \frac{M_u \cdot (l_{w.tot}/2)}{I_g} = 0.502 \text{ ksi}.$$

Check if boundary elements are required:

$$f_c = 0.502 \text{ ksi} \ll 0.2 f_{cp} = 2 \text{ ksi}.$$

Therefore, boundary elements are not required.

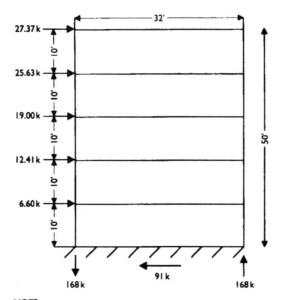

NOTE:
LOADS ARE CALCULATED FROM A FIVE STORY BUILDING 120' LONG ×70' WIDE WITH 8"
REINFORCED CONCRETE FLOORS AND ROOF SLABS. CALCULATION FOR LOADS AND LOAD
DISTRIBUTION ARE BEYOND THE SCOPE OF THIS EXAMPLE.

Figure 5.28 Applied service loads at shear wall.

Calculate the required web reinforcement, $V_u = 127.4$ kips.
Shear area:

$$A_{cv} = (l_w + l_b) \cdot t_w = 2.304 \times 10^3 \text{ in}^2.$$

$$V_u = 127.4 \text{ kips} \ll 2 \cdot A_{cv} \cdot \sqrt{f_{cp} \text{ (psi)}} = 460.8 \text{ kips}$$

One curtain of reinforcement is required. Minimum longitudinal and transverse
reinforcement ratio = 0.0025. Maximum bar spacing = 18 in.

Calculate nominal shear strength:
Calculate α_c:

$$\frac{h_w}{l_w + l_b} = 1.563 \le 2$$

$$\alpha_c = \text{if} \left[\frac{h_w}{l_w + l_b} \le 1.5, 3.0, \text{if} \left[\frac{h_w}{l_w + l_b} \ge 2.0, 2.0, 6.0 \right. \right.$$
$$\left. \left. -2 \cdot \left(\frac{h_w}{l_w + l_b} \right) \right] \right] = 2.875.$$

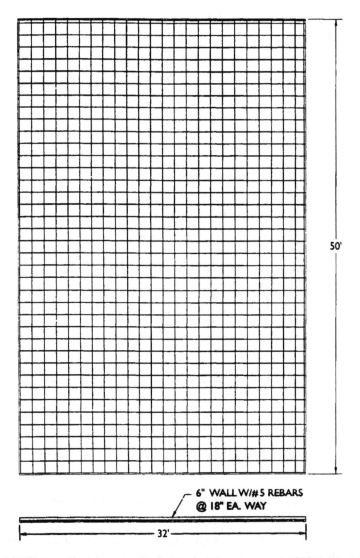

Figure 5.29 Shear wall reinforcement in section and elevation as per ACI 318-95 guidelines.

$$A_{cv} = (l_w + l_b)t_w = 2.304 \times 10^3 \text{ in}^2.$$

Assume $N_{wb} = 1\#$, $X_{wb} = 5$ at $s_{wb} = 18$ in.

$$A_{rb} = A_{b_{x_{wb}}} \cdot \frac{12 \text{ in}}{s_{wb}}, \qquad A_{wb} = A_{rb} \cdot N_{wb} = 0.207 \text{ in}^2,$$

$$\rho_{wb} = \frac{A_{wb}}{t_w \cdot 12 \text{ in}} = 2.87 \times 10^{-3},$$

$$\rho_n = \text{if } (\rho_{wb} \leq 0.0025, 0.0025, \rho_{wb}) = 2.87 \times 10^{-3}.$$

Calculate allowable shear load, $\phi = 0.6$.

$$\phi V_n = \phi \cdot A_{cv} \cdot \left[\alpha_c \cdot \sqrt{f_{cp}(\text{psi})} + \rho_n \cdot f_y \right]$$

$$= 635.52 \text{ kips} \gg V_u = 127.4 \text{ kips (OK)}.$$

Use 1–#5 at 18 in o/c in the horizontal direction.
Vertical distribution of reinforcement:

Use minimum $\rho_v = \text{if } \left(\left(\frac{hw}{7} l_w + l_b \right) \leq 2, \rho_n, 0.0025 \right) = 2.87 \times 10^{-3}$.

$$A_{sv} = \rho_v \cdot t_w = 0.207 \text{ in}^2/\text{ft}.$$

For $N_{wb} = 1\#$, $x_{wb} = 5$ at $A_{b_{x_{wb}}} = 0.31 \text{ in}^2$,

$$S = \frac{A_{b_{x_{wb}}}}{A_{sv}} = 18 \text{ in} \gg s_{wb} = 18 \text{ in (OK)}.$$

Use 1–#5 at 18 in o/c in the vertical direction.

5.11.6.2 Solution (2a): triangular reinforcement method where the ratio of width to height is smaller than one

For this method, the wall is assumed to consist of members of truss elements as given below. Each member is 6 in thick (the required thickness of the wall) and has a certain depth as determined by the following calculation for the most critically loaded member. A two-dimensional, finite element frame analysis program was used to calculate loads on members of truss. From the analysis, the most critically loaded member is member 19–21 (see Member End Forces of analysis). By designing this member using an RC design program, a 6 in × 26 in section is found to be adequate. Therefore, a truss 6 in thick ×26 in deep members with 4 #6 rebars and confinement as shown is adequate to resist applied lateral loads. Calculation for vertical beams supporting vertical live and dead floor and roof loads is beyond

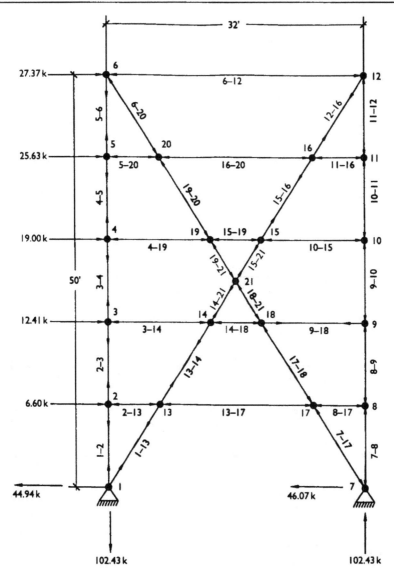

Figure 5.30 Frame members and applied lateral service loads using the triangular reinforcement method. Spaces between truss members will be void with no concrete casting.

the scope of this example. This example is given to show the feasibility of using the newly proposed triangular reinforcement method in comparison with the traditional shear wall design method outlined in the ACI guidelines (Figure 5.30–5.32 and Tables 5.5 and 5.6).

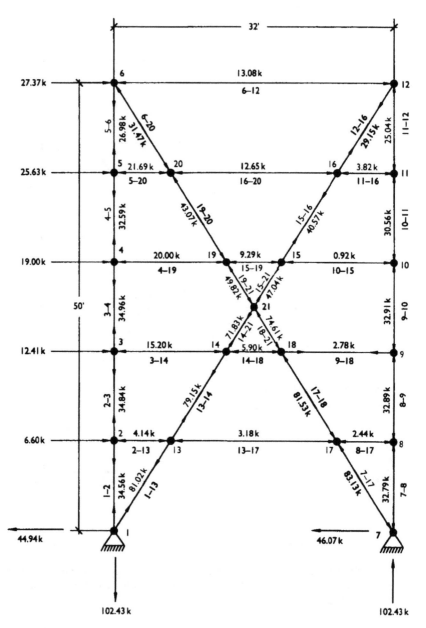

Figure 5.31 Axial forces, reactions and applied loads. See results of the two-dimensional frame analysis program for other shear and moment loads.

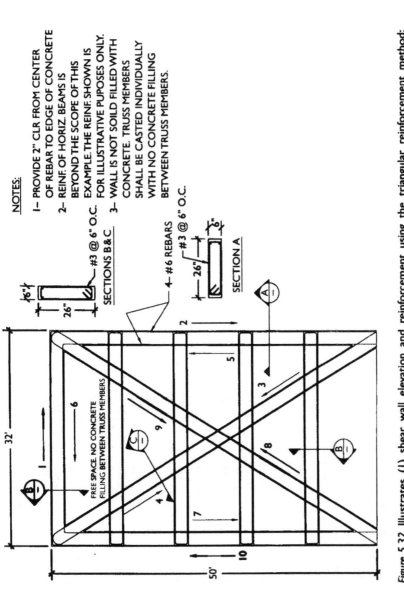

NOTES:

1– PROVIDE 2" CLR FROM CENTER OF REBAR TO EDGE OF CONCRETE

2– REINF. OF HORIZ. BEAMS IS BEYOND THE SCOPE OF THIS EXAMPLE. THE REINF. SHOWN IS FOR ILLUSTRATIVE PUPOSES ONLY.

3– WALL IS NOT SOILD FILLED WITH CONCRETE. TRUSS MEMBERS SHALL BE CASTED INDIVIDUALLY WITH NO CONCRETE FILLING BETWEEN TRUSS MEMBERS.

#3 @ 6" O.C.

SECTIONS B & C

4 – #6 REBARS

#3 @ 6" O.C.

26"

6"

SECTION A

FREE SPACE. NO CONCRETE FILLING BETWEEN TRUSS MEMBERS

32'

50'

26"

6"

Figure 5.32 Illustrates (1) shear wall elevation and reinforcement using the triangular reinforcement method; (2) directional path for using only one rebar each side of the wall; (3) typical sections showing reinforcement and confinement.

Table 5.5 Two-dimensional frame analysis

Nodes . . .

Node label	Node coordinates		X restraint	Y restraint	Z restraint	Node temp (deg F)
	X (ft)	Y (ft)				
1	0.000	0.000	Fixed	Fixed		0
2	0.000	10.000				0
3	0.000	20.000				0
4	0.000	30.000				0
5	0.000	40.000				0
6	0.000	50.000				0
7	32.000	0.000	Fixed	Fixed		0
8	32.000	10.000				0
9	32.000	20.000				0
10	32.000	30.000				0
11	32.000	40.000				0
12	32.000	50.000				0
13	6.400	10.000				0
14	12.800	20.000				0
15	19.200	30.000				0
16	25.600	40.000				0
17	25.600	10.000				0
18	19.200	20.000				0
19	12.800	30.000				0
20	6.400	40.000				0
21	16.000	25.000				0

Member ...

Member label	Property label	Endpoint nodes		Member length (ft)	I end releases			J end releases		
		I node	J node		X	Y	Z	X	Y	Z
1–13	Memb	1	13	11.873			Free			
1–2	col	1	2	10.000						
10–11	col	10	11	10.000						
10–15	Hriz.bm	10	15	12.800			Free			
11–12	col	11	12	10.000						
11–16	Hriz.bm	11	16	6.400			Free			
12–16	Memb	12	16	11.873			Free			
13–14	Memb	13	14	11.873						
13–17	Hriz.bm	13	17	19.200						
14–18	Hriz.bm	14	18	6.400						
14–21	Memb	14	21	5.936						
15–16	Memb	15	16	11.873						
15–19	Hriz.bm	15	19	6.400						
15–21	Memb	15	21	5.936						
16–20	Hriz.bm	16	20	19.200						
17–18	Memb	17	18	11.873						
18–21	Memb	18	21	5.936						
19–20	Memb	19	20	11.873						
19–21	Memb	19	21	5.936						
2–13	Hriz.bm	2	13	6.400			Free			
2–3	col	2	3	10.000						
3–14	Hriz.bm	3	14	12.800			Free			
3–4	col	3	4	10.000						
4–19	Hriz.bm	4	19	12.800			Free			
4–5	col	4	5	10.000						
5–20	Hriz.bm	5	20	6.400			Free			

Table 5.5 (Continued)

Member ...

Member label	Property label	Endpoint nodes		Member length (ft)	I end releases			J end releases		
		I node	J node		X	Y	Z	X	Y	Z
5-6	col	5	6	10.000						Free
6-12	Hriz.bm	6	12	32.000			Free			
6-20	Memb	6	20	11.873			Free			
7-17	Memb	7	17	11.873			Free			
7-8	col	7	8	10.000						
8-17	Hriz.bm	8	17	6.400			Free			
8-9	col	8	9	10.000						
9-10	col	9	10	10.000						
9-18	Hriz.bm	9	18	12.800			Free			

Materials ...

Member label	Youngs (ksi)	Density (kcf)	Thermal (in/100d)	Yield (ksi)
Concrete	5,700.00	0.145	0.000000	60.00
Default	1.00	0.000	0.000000	1.00
Steel	29,000.00	0.490	0.000650	36.00

Section sections . . .

Prop label Group tag	Material	Area	Depth width	Tf Tw	Ixx Iyy
6 × 26 Hriz_bm	Concrete	156.000 in²	26.000 in 6.000 in	0.000 in 0.000 in	8,788.00 in⁴ 0.00 in⁴
6 × 26 Memb	Concrete	156.000 in²	26.000 in 6.000 in	0.000 in 0.000 in	8,788.00 in⁴ 0.00 in⁴
6 × 26 col	Concrete	156.000 in²	26.000 in 6.000 in	0.000 in 0.000 in	8,788.00 in⁴ 0.00 in⁴
Default	Default	1.000 in²	0.000 in 0.000 in	0.000 in 0.000 in	1.00 in⁴ 0.00 in⁴

Node loads . . .

Node label	Concentrated loads and moments			Load case factors				
	X	Y	Moment	#1	#2	#3	#4	#5
2	6.600 k			1.000				
3	12.410 k			1.000				
4	19.000 k			1.000				
5	25.630 k			1.000				
6	27.370 k			1.000				

Load combinations . . .

Load combination description	Stress increase	Gravity load factors		Load combination factors				
		X	Y	#1	#2	#3	#4	#5
Lateral	1.000			1.000				

Table 5.5 (Continued)
Node displacements and reactions

Node label	Load combination	Node displacements			Node reactions		
		X (in)	Y (in)	Z (Radians)	X (k)	Y (k)	Z (k ft)
1	Lateral	0	0	-0.00028	-44.94181	-102.43437	0
2	Lateral	0.02983	0.00466	-0.00018	0	0	0
3	Lateral	0.05009	0.00937	-0.00024	0	0	0
4	Lateral	0.09173	0.01408	-0.00041	0	0	0
5	Lateral	0.13003	0.01848	-0.00014	0	0	0
6	Lateral	0.12416	0.02212	0.00014	0	0	0
7	Lateral	0	0	-0.00027	-46.06818	102.43437	0
8	Lateral	0.02844	-0.00442	-0.00017	0	0	0
9	Lateral	0.04744	-0.00886	-0.00023	0	0	0
10	Lateral	0.08764	-0.01330	-0.00039	0	0	0
11	Lateral	0.12455	-0.01743	-0.00013	0	0	0
12	Lateral	0.11851	-0.02081	0.00014	0	0	0
13	Lateral	0.02947	-0.00345	-0.00009	0	0	0
14	Lateral	0.04747	0.00009	-0.00004	0	0	0
15	Lateral	0.08748	-0.01421	-0.00036	0	0	0
16	Lateral	0.12488	-0.03043	-0.00005	0	0	0
17	Lateral	0.02865	0.00252	-0.00009	0	0	0
18	Lateral	0.04696	-0.00127	-0.00005	0	0	0
19	Lateral	0.08828	0.01334	-0.00038	0	0	0
20	Lateral	0.12816	0.03067	-0.00006	0	0	0
21	Lateral	0.05916	-0.00056	-0.00034	0	0	0

Member end forces ...

Member label	Load combination	Node "i" end forces			Node "j" end forces		
		Axial (k)	Shear (k)	Moment (ft k)	Axial (k)	Shear (k)	Moment (ft k)
1–13	Lateral	−81.02409	0.68686	0	81.02409	−0.68686	8.15488
1–2	Lateral	−34.56036	0.68696	0	34.56036	−0.68696	6.86958
10–11	Lateral	30.55864	1.94099	0.58614	−30.55864	−1.94099	18.82376
10–15	Lateral	−0.92254	−2.35467	0	0.92254	2.35467	−30.13972
11–12	Lateral	25.04143	−1.88238	−18.82376	−25.04143	1.88238	0
11–16	Lateral	3.82336	−5.51722	0	−3.82336	5.51722	−35.31018
12–16	Lateral	−29.15603	−0.89812	0	29.15603	0.89812	−10.66312
13–14	Lateral	−79.15471	0.73654	2.82670	79.15471	−0.73654	5.91805
13–17	Lateral	3.17618	−1.31770	−12.79664	−3.17618	1.31770	−12.50313
14–18	Lateral	5.90895	−2.63797	−8.23850	−5.90895	2.63797	−8.64454
14–21	Lateral	−71.82625	−5.60026	0.80103	71.82625	5.60026	−34.04598
15–16	Lateral	−40.57294	2.27968	4.25880	40.57294	−2.27968	22.80706
15–19	Lateral	9.29013	−1.21129	−3.13994	−9.29013	1.21129	−4.61231
15–21	Lateral	−47.04115	10.26518	29.02086	47.04115	−10.26518	31.91661
16–20	Lateral	12.65426	2.38591	23.16624	−12.65426	−2.38591	22.64327
17–18	Lateral	81.53499	0.82759	3.71394	−81.53499	−0.82759	6.11176
18–21	Lateral	74.61071	−5.05132	2.21149	−74.61071	5.05132	−32.19777
19–20	Lateral	43.07671	2.42320	5.13680	−43.07671	−2.42320	23.63303
19–21	Lateral	49.82291	10.80991	29.84402	−49.82291	−10.80991	34.32714
2–13	Lateral	4.14203	0.28360	0	−4.14203	−0.28360	1.81507
2–3	Lateral	−34.84396	−1.77102	−6.86958	34.84396	1.77102	−10.84058
3–14	Lateral	15.19669	0.11871	0	−15.19669	−0.11871	1.51943
3–4	Lateral	−34.96267	1.01568	10.84058	34.96267	−1.01568	−0.68381
4–19	Lateral	19.99058	−2.37254	0	−19.99058	2.37254	−30.36851
4–5	Lateral	−32.59013	2.00626	0.68381	32.59013	−2.00626	19.37878
5–20	Lateral	21.68586	−5.61458	0	−21.68586	5.61458	−35.93333

Table 5.5 (Continued)
Member end forces . . .

Member label	Load combination	Node "i" end forces			Node "j" end forces		
		Axial (k)	Shear (k)	Moment (ft k)	Axial (k)	Shear (k)	Moment (ft k)
5–6	Lateral	−26.97555	−1.93788	−19.37878	26.97555	1.93788	0
6–12	Lateral	13.07783	0	0	−13.07783	0	0
6–20	Lateral	31.46960	−0.87116	0	−31.46960	0.87116	−10.34297
7–17	Lateral	83.13021	0.68484	0	−83.13021	−0.68484	8.13081
7–8	Lateral	32.78534	0.67972	0	−32.78534	−0.67972	6.79719
8–17	Lateral	2.43651	0.10287	0	−2.43651	−0.10287	0.65837
8–9	Lateral	32.88821	−1.75679	−6.79719	−32.88821	1.75679	−10.77068
9–10	Lateral	32.91331	1.01845	10.77068	−32.91331	−1.01845	−0.58614
9–18	Lateral	−2.77524	0.02510	0	2.77524	−0.02510	0.32129

Member overall envelope summary

Member label	Section	Axial (k)	Shear (k)	Moment (ft k)	Deflection (in)
1–13	Memb	81.024	0.687	8.155	0.026
1–2	col	34.560	0.687	6.870	0.030
10–11	col	30.559	1.941	18.824	0.037
10–15	Hriz_bm	0.923	2.355	30.140	0.012
11–12	col	25.041	1.882	18.824	0.005
11–16	Hriz_bm	3.823	5.517	35.310	0.013
12–16	Memb	29.156	0.898	10.663	0.010
13–14	Memb	79.155	0.737	5.918	0.013
13–17	Hriz_bm	3.176	1.318	12.797	0.007
14–18	Hriz_bm	5.909	2.638	8.645	0.001

14–21	Memb	71.826	5.600	34.046	0.010
15–16	Memb	40.573	2.280	22.807	0.041
15–19	Hriz_bm	9.290	1.211	4.612	0.028
15–21	Memb	47.041	10.265	31.917	0.031
16–20	Hriz_bm	12.654	2.386	23.166	0.061
17–18	Memb	81.535	0.828	6.112	0.013
18–21	Memb	74.611	5.051	32.198	0.010
19–20	Memb	43.077	2.423	23.633	0.044
19–21	Memb	49.823	10.810	34.327	0.032
2–13	Hriz_bm	4.142	0.284	1.815	0.008
2–3	col	34.844	1.771	10.841	0.020
3–14	Hriz_bm	15.197	0.119	1.519	0.009
3–4	col	34.963	1.016	10.841	0.041
4–19	Hriz_bm	19.991	2.373	30.369	0.012
4–5	col	32.590	2.006	19.379	0.039
5–20	Hriz_bm	21.686	5.615	35.933	0.012
5–6	col	26.976	1.938	19.379	0.005
6–12	Hriz_bm	13.078			0.042
6–20	Memb	31.470	0.871	10.343	0.008
7–17	Memb	83.130	0.685	8.131	0.025
7–8	col	32.785	0.680	6.797	0.029
8–17	Hriz_bm	2.437	0.103	0.658	0.007
8–9	col	32.888	1.757	10.771	0.019
9–10	col	32.913	1.018	10.771	0.040
9–18	Hriz_bm	2.775	0.025	0.321	0.008

Deflection values listed are the maximum of a sampling of 31 points across the member.

Table 5.6 Rectangular concrete column – description, 5 story building, member 19–21

General information		Calculations are designed to ACI 318-95 and 1997 UBC requirements		
Width	6.000 in	f_c	10,000.0 psi	Total height 0.000 ft
Depth	26.000 in	F_y	60,000.0 psi	Unbraced length 12.000 ft
Rebar:		Seismic zone	4	Eff. length factor
2-#6 d = 2.000 in		LL & ST loads act together		Column is BRACED 1.000
2-#6 d = 24.000 in				

Loads

	Dead load	Live load	Short term	Eccentricity (in)
Axial loads	49.823 k	69.75 k (k)	44.84 k (k)	
Applied moments ...				
@ Top	29.844 k ft	(k ft)	(k ft)	
@ Bottom	34.327 k ft	(k ft)	(k ft)	

Summary — Column is OK

6.00 × 26.00 in column, rebar: 2-#6 @ 2.00 in, 2-#6 @ 24.00 in

	ACI 9-1	ACI 9-2	ACI 9-3
Applied: P_u: Max factored	69.75 k	69.75 k	44.84 k
Allowable: P_n * Phi @ Design Ecc.	186.44 k	186.44 k	85.68 k
M-critical	40.67 k ft	40.67 k ft	40.67 k ft
Combined eccentricity	6.997 in	6.997 in	10.884 in
Magnification factor	1.00	1.00	1.00
Design eccentricity	14.659 in	14.659 in	22.802 in
Magnified design moment	40.67 k ft	40.67 k ft	40.67 k ft
P_o * .80	1,133.31 k	1,133.31 k	1,133.31 k
P : Balanced	354.35 k	354.35 k	354.35 k
Ecc : Balanced	11.620 in	11.620 in	11.620 in

Slenderness (per ACI 318-95 section 10.12 & 10.13)

Actual k L_u/r 18.462	Elastic modulus	5,700.0 ksi	Beta	0.650
	ACI Eq. 9-1	ACI Eq. 9-2	ACI Eq. 9-3	
Neutral axis distance	8.2600 in	8.2600 in	3.9050 in	
Phi	0.7000	0.7000	0.8044	
Max limit kl/r	44.4328	44.4328	44.4328	
Beta = M:sustained/M:max	1.0000	1.0000	1.0000	
Cm	1.0000	1.0000	1.0000	
EI/1000	0.00	0.00	0.00	
P_c : pi^2 E I/(k L_u)^2	0.00	0.00	0.00	
alpha: MaxP_u/(.75 P_c)	0.0000	0.0000	0.0000	
Delta	1.0000	1.0000	1.0000	
Ecc: Ecc Loads + Moments	6.9968	6.9968	10.8839 in	
Design Ecc = Ecc * Delta	14.6586	14.6586	22.8022 in	

ACI factors (per ACI, applied internally to entered loads)

ACI 9-1 & 9-2 DL	1.400	ACI 9-2 group factor	0.750	UBC 1921.2.7 "1.4" factor	1.400
ACI 9-1 & 9-2 LL	1.700	ACI 9-3 dead load factor	0.900	UBC 1921.2.7 "0.9" factor	0.900
ACI 9-1 & 9-2 ST	1.700	ACI 9-3 short term factor	1.300	
...seismic = ST*	1.100			

5.11.6.3   Solution (2b): triangular reinforcement method where the ratio of width to height is larger than one

For this method, the wall is assumed to consist of members of truss elements as shown below. Each member is 6 in thick (the required thickness of the wall) and has a certain depth as determined by the following calculation for the most critically loaded member. A two-dimensional, finite-element frame analysis program was used to calculate loads on members of truss. From the analysis, the most critically loaded member is member 3–7 (see Member End Forces of analysis). By designing this member using an RC design program, a 6 in × 24 in section is found to be adequate. Therefore, a truss 6 in thick ×26 in deep members with 4 #6 rebars and confinement as shown is adequate to resist applied lateral loads. Calculation for vertical beams supporting vertical live and dead floor and roof loads is beyond the scope of this example. This example is given to show the feasibility of using the newly proposed triangular reinforcement method in comparison with the traditional shear wall design method outlined in the ACI guidelines (Figures 5.33–5.35 and Tables 5.7 and 5.8).

5.11.6.4   Brief comments

For a shear wall 32 ft wide and 50 ft high, by applying the classical concept, it appears that we need a total reinforcement of 2,572 lb (see Section 5.11.6, Figure 5.29). For the same shear wall, by applying the new concept, we need 4,115 lb of reinforcement (see Section 5.11.5.3, Figure 5.35). Comparing panel X-1 with panel C-1 (see Section 5.11.3), we see that the new system is two times stronger than the old one, meaning that the classical order concept needs two times more reinforcement to reach the same resistance as the new system. This means that we need 2,572 × 2 = 5,144 lb of total reinforcement. By comparing this reinforcement with the reinforcement for the new system (Figure 5.35), it follows that the new system (triangular reinforcement) is cheaper by 25 percent.

In fact, the economical comparison above is irrelevant because the old system is based on the diagonal tension theory, developed by Ritter and Morsch, which, as such, is not accurate. At the same time, the new design is based on a new law of physics where reinforcement reacts to external load as soon as such a load is applied. On the other hand, in the classical approach, reinforcement reacts (horizontally or vertically) only when such a reinforcement is crossed by the cracking of the concrete.

Furthermore, the essential concept of rigidity, necessary for any shear wall, is ensured by triangular reinforcement, while such rigidity in a classical shear wall simply does not exist because its deformation is a function of the strength of the concrete and not the reinforcement.

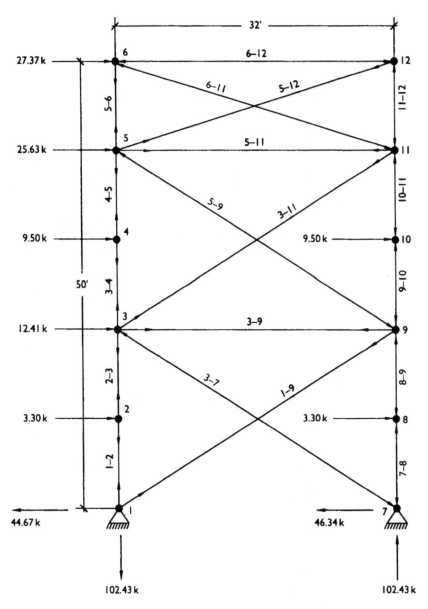

Figure 5.33 Frame members and applied lateral service loads using the triangular reinforcement method. Spaces between truss members enclosed by nodes 3, 5, 9 and 11 will be void with no concrete casting.

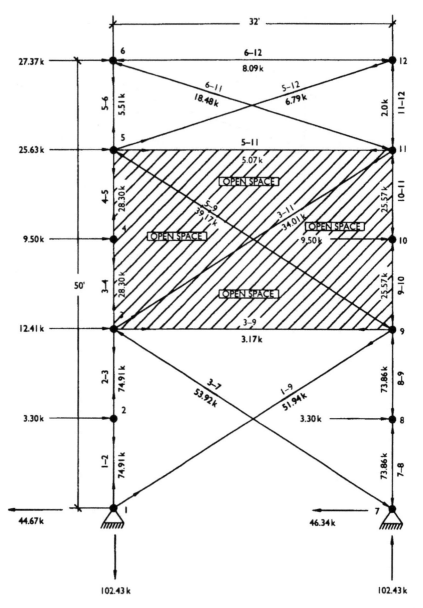

Figure 5.34 Axial forces, reactions and applied loads. See results of the two-dimensional frame analysis program for other shear and moment loads.

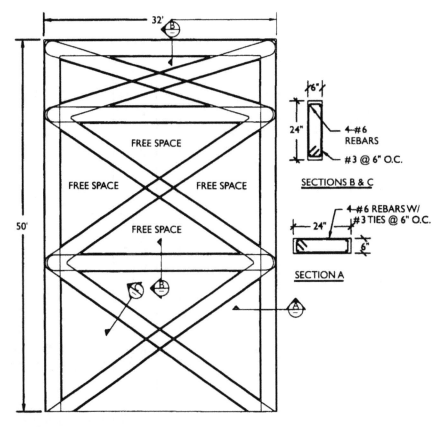

SECTIONS B & C

SECTION A

NOTES:
1. PROVIDE 2" CLR FROM CENTER OF REBAR TO EDGE OF CONCRETE
2. REINF. OF HORIZ. BEAMS IS BEYOND THE SCOPE OF THIS EXAMPLE.
   THE REINF. SHOWN IS FOR ILLUSTRATIVE PUPOSES ONLY.
3. WALL IS NOT SOILD FILLED WITH CONCRETE WHERE "FREE SPACE"
   NOTE IS SHOWN.

Figure 5.35 (1) Illustrates shear wall elevation and reinforcement using the triangular rein-
            forcement method; (2) directional path for using only one rebar each side of
            the wall; (3) typical sections showing reinforcement and confinement.

Table 5.7 Two-dimensional frame analysis

Nodes...

Node label	Node coordinates		X restraint	Y restraint	Z restraint	Node temp (deg F)
	X (ft)	Y (ft)				
1	0.000	0.000	Fixed	Fixed		0
2	0.000	10.000				0
3	0.000	20.000				0
4	0.000	30.000				0
5	0.000	40.000				0
6	0.000	50.000				0
7	32.000	0.000	Fixed	Fixed		0
8	32.000	10.000				0
9	32.000	20.000				0
10	32.000	30.000				0
11	32.000	40.000				0
12	32.000	50.000				0

Member...

Member label	Property label	Endpoint nodes		Member length (ft)	I end releases			J end releases		
		I node	J node		X	Y	Z	X	Y	Z
1-2	col	1	2	10.000						
1-9	Memb	1	9	37.736			Free			Free
10-11	col	10	11	10.000						
11-12	col	11	12	10.000						
2-3	col	2	3	10.000						
3-11	Memb	3	11	37.736			Free			Free
3-4	col	3	4	10.000						
3-7	Memb	3	7	37.736			Free			Free

Member label	Type	i	j	Value		
3–9	Hriz_bm	3	9	32.000	Free	Free
4–5	col	4	5	10.000		
5–11	Hriz_bm	5	11	32.000	Free	Free
5–12	Memb	5	12	33.526	Free	Free
5–6	col	5	6	10.000		
5–9	Memb	5	9	37.736	Free	Free
6–11	Memb	6	11	33.526	Free	Free
6–12	Hriz_bm	6	12	32.000	Free	Free
7–8	col	7	8	10.000		
8–9	col	8	9	10.000		
9–10	col	9	10	10.000		

Materials

Member label	Youngs (ksi)	Density (kcf)	Thermal (in/100d)	Yield (ksi)
Concrete	5,700.00	0.145	0.000000	60.00
Default	1.00	0.000	0.000000	1.00
Steel	29,9000.00	0.490	0.000650	36.00

Section sections. . .

Prop label / Group tag	Material	Area	Depth / Width	Tf / Tw	Ixx / Iyy
6 × 24 / col	Concrete	144.000 in²	24.000 in / 6.000 in	0.000 in / 0.000 in	6,912.00 in⁴ / 0.00 in⁴
6 × 24 / Memb	Concrete	144.000 in²	24.000 in / 6.000 in	0.000 in / 0.000 in	6,912.00 in⁴ / 0.00 in⁴
6 × 24 / Hriz_bm	Concrete	144.000 in²	24.000 in / 6.000 in	0.000 in / 0.000 in	6,912.00 in⁴ / 0.00 in⁴
Default	Default	1.000 in²	0.000 in / 0.000 in	0.000 in / 0.000 in	1.00 in⁴ / 0.00 in⁴

Table 5.7 (Continued)

Node loads...

Node Label	Concentrated loads and moments			Load case factors				
	X	Y	Moment	#1	#2	#3	#4	#5
2	3.300k			1.000				
3	12.410k			1.000				
4	9.500k			1.000				
5	25.630k			1.000				
6	27.370k			1.000				
8	3.300k			1.000				
10	9.500k			1.000				

Load combinations ...

Load combination description	Stress increase	Gravity load factors		Load combination factors				
		X	Y	#1	#2	#3	#4	#5
Lateral	1.000			1.000				

Node displacements & reactions

Node label	Load combination	Node displacements			Node reactions		
		X (in)	Y (in)	Z (Radians)	X (k)	Y (k)	Z (k ft)
1	Lateral	0	0	−0.00025	−44.66767	−102.43437	0
2	Lateral	0.02597	0.01095	−0.00014	0	0	0
3	Lateral	0.04877	0.02190	−0.00040	0	0	0
4	Lateral	0.10590	0.02604	−0.00025	0	0	0
5	Lateral	0.10513	0.03018	−0.00002	0	0	0

6	Lateral	0.13101	0.03098	-0.00032	0	-46.34233	0
7	Lateral	0	0	-0.00025	0	0	102.43437
8	Lateral	0.02510	-0.01080	-0.00013	0	0	0
9	Lateral	0.04729	-0.02160	-0.00040	0	0	0
10	Lateral	0.10413	-0.02533	-0.00024	0	0	0
11	Lateral	0.10276	-0.02907	-0.00001	0	0	0
12	Lateral	0.12723	-0.02937	-0.00030	0	0	0

Member end forces ...

Member label	Load combination	Node "i" end forces Axial (k)	Shear (k)	Moment (ft·k)	Node "j" end forces Axial (k)	Shear (k)	Moment (ft·k)
1-2	Lateral	-74.90597	0.62222	0	74.90597	-0.62222	6.22223
1-9	Lateral	-51.94049	0	0	51.94049	0	0
10-11	Lateral	25.56593	-4.52223	-29.10707	-25.56593	4.52223	-16.11526
11-12	Lateral	2.02602	1.61153	16.11526	-2.02602	-1.61153	0
2-3	Lateral	-74.90597	-2.67778	-6.22223	74.90597	2.67778	-20.55554
3-11	Lateral	-34.01535	0	0	34.01535	0	0
3-4	Lateral	-28.29921	4.95907	20.55554	28.29921	-4.95907	29.03515
3-7	Lateral	53.92211	0	0	-53.92211	0	0
3-9	Lateral	3.16594	0	0	-3.16594	0	0
4-5	Lateral	-28.29921	-4.54093	-29.03515	28.29921	4.54093	-16.37416
5-11	Lateral	5.07340	0	0	5.07340	0	0
5-12	Lateral	-6.79245	0	0	6.79245	0	0
5-6	Lateral	-5.51181	1.63742	16.37416	5.51181	-1.63742	0
5-9	Lateral	39.17249	0	0	-39.17249	0	0
6-11	Lateral	18.47896	0	0	-18.47896	0	0
6-12	Lateral	8.09478	0	0	-8.09478	0	0
7-8	Lateral	73.85571	0.61647	0	-73.85571	-0.61647	6.16469
8-9	Lateral	73.85571	-2.68353	-6.16469	-73.85571	2.68353	-20.67061
9-10	Lateral	25.56593	4.97777	20.67061	-25.56593	-4.97777	29.10707

Table 5.7 (Continued)

Member overall envelope summary

Member label	Section	Axial (k)	Shear (k)	Moment (ft k)	Deflection (in)	Maximum stress ratio
1–2	col	74.906	0.622	6.222	0.026	
1–9	Memb	51.940			0.042	
10–11	col	25.566	4.522	29.107	0.004	
11–12	col	2.026	1.612	16.115	0.024	
2–3	col	74.906	2.678	20.556	0.022	
3–11	Memb	34.015			0.069	
3–4	col	28.299	4.959	29.035	0.057	
3–7	Memb	53.922			0.043	
3–9	Hriz_bm	3.166			0.042	
4–5	col	28.299	4.541	29.035	0.005	
5–11	Hriz_bm	5.073			0.057	
5–12	Memb	6.792			0.061	
5–6	col	5.512	1.637	16.374	0.025	
5–9	Memb	39.172			0.072	
6–11	Memb	18.479			0.064	
6–12	Hriz_bm	8.095			0.058	
7–8	col	73.856	0.616	6.165	0.025	
8–9	col	73.856	2.684	20.671	0.022	
9–10	col	25.566	4.978	29.107	0.057	

Deflection values listed are the maximum of a sampling of 31 points across the member.

Table 5.8 Rectangular concrete column – description, member 3–7

General information

Calculations are designed to ACI 318-95 and 1997 UBC requirements

Width	6.000 in	fc	10,000.0 psi	Total height	35.000 ft
Depth	24.000 in	Fy	60,000.0 psi	Unbraced length	35.000 ft
Rebar:		Seismic zone	4	Eff. length factor	1.000
2–#6d = 2.000 in		LL & ST loads act separately		Column is BRACED	
2–#6d = 24.000 in					

Loads

Axial loads	Dead load (k)	Live load (k)	Short term (k)	Eccentricity (in)
	1.000 k		54.000 k	

Summary

6.00 × 24.00 in column, rebar: 2–#6 @ 2.00 in, 2–#6 @ 22.00 in   Column is OK

	Dead load	Live load	Short term	
	ACI 9-1	ACI 9-2	ACI 9-3	
Applied: P_u: Max Factored	1.40 k	77.00 k	76.50 k	
Allowable: P_n * Phi @ Design Ecc.	1.73 k	163.93 k	162.07 k	
M-critical	40.67 k ft	40.67 k ft	40.67 k ft	
Combined Eccentricity	348.600 in	6.338 in	6.380 in	
Magnification Factor	1.00	1.13	1.13	
Design Eccentricity	731.817 in	14.131 in	14.211 in	
Magnified Design Moment	40.84 k ft	46.14 k ft	46.06 k ft	
P_o * .80	1,051.71 k	1,051.71 k	1,051.71 k	
P : Balanced	324.35 k	324.35 k	324.35 k	
Ecc : Balanced	10.985 in	10.985 in	10.985 in	

Table 5.8 (Continued)

Slenderness (per ACI 318-95 Sections 10.12 and 10.13)

Actual k L$_u$/r	58.333		Elastic modulus	5,700.0 ksi	Beta	0.650
			ACI Eq. 9-1	ACI Eq. 9-2	ACI Eq. 9-3	
Neutral axis distance			1.8450 in	7.2900 in	7.2100 in	
Phi			0.8981	0.7000	0.7000	
Max limit kl/r			34.0000	34.0000	34.0000	
Beta = M : sustained/M : max			1.0000	0.0182	0.0118	
C$_m$			1.0000	1.0000	1.0000	
EI/1000			7,879.68	15,477.94	15,576.11	
P$_c$:pi^2 E I/(k L$_u$)^2			440.87	865.99	871.49	
alpha: MaxP$_u$/(.75 P$_c$)			0.0042	0.1186	0.1170	
Delta			1.0043	1.1345	1.1326	
Ecc: Ecc loads + moments			348.6000	6.3382	6.3796 in	
Design Ecc = Ecc * Delta			731.8167	14.1313	14.2113 in	

ACI Factors (per ACI, applied internally to entered loads)

ACI 9-1 & 9-2 DL	1.400	ACI 9-2 group factor	0.750	UBC 1921.2.7 "1.4" factor	1.400
ACI 9-1 & 9-2 LL	1.700	ACI 9-3 dead load factor	0.900	UBC 1921.2.7 "0.9" factor	0.900
ACI 9-1 & 9-2 ST	1.700	ACI 9-3 short term factor	1.300		
. . . seismic = ST' :	1.100				

Yet, the new system possesses inherent ductility so that rupture can occur only after exhaustion of the plastic stage of reinforcement, while the classical shear wall ruptured as soon as concrete cracking passed through the shear wall.

We make special note of the fact that, in the new approach, there is no necessity to avoid a cantilevered beam (with neutral axis in the body), as is illustrated in Section 5.11.6.3, Figure 5.34.

Finally, we note that designing a shear wall as a cantilever beam, where the neutral axis is developed (height is larger than width) and where the inside of such a beam is treated as a compressive strut in the tensile zone, is simply a utopian concept – as is perpetual motion without replenished energy. So, if the laws of physics are respected, the following question arises: how could a compressive strut ever be developed in the tensile zone of a cantilevered beam?

Acknowledgments

Donation of concrete: As his contribution to this research, Mr. William Miller, manager of 'Concrete Products Incorporated', of Corona, California, donated the concrete, the casting of concrete, curing 28 days and the testing of concrete cylinders, for all panels. For his contribution to this project, the author is deeply thankful to Mr. Miller and his company.

Donation of testing: Smith-Emery Company from Los Angeles, California carried out the testing for all panels, for which the author is very deeply thankful, specifically to the brilliant manager of the laboratory, Mr. Rico Lubas.

Calculations: Computer calculations were performed by the Koujah Consulting Engineering Firm, Riverside, California.

This author is forever grateful to the Construction Company of Mr. Nick Tavaglione from Riverside, California for his financial support for shear wall panel preparation.

References

1. Park, R. and Paulay, T. (1975) *Reinforced Concrete Structures*, John Wiley and Sons, New York, pp. 270, 271, 307, 308, Figure 7.20, 611, 613, 618.
2. Barda, F., Hanson, J. M. and Corley, W. G. (1978) "Shear Strength of Low-Rise Walls with Boundary Elements", *Reinforced Concrete Structures*, ACI Publication SP-53, 2nd printing, Detroit, MI, pp. 150–151, 154.
3. Berg, V. B. and Stratta, J. L. (1964) "Anchorage and the Alaska Earthquake of March 27, 1964", American Iron and Steel Institute, New York, p. 66.
4. Anderson, B. G. (1957) "Rigid Frame Failure", *ACI Journal* 28(7), 625–636.
5. McGuire, W. (1959) "Reinforced Concrete", *Civil Engineering Handbook*, 4th edn, editor-in-chief, Leonard Church Urquhart, McGraw-Hill Book Company, New York, pp. 7–129.
6. Esteva, L. (1980) "Design: General", *Design of Earthquake Resistant Structures*, ed E. Rosenblueth, John Wiley and Sons, New York, Chapter 3.

7. Bertero, V. V. and Popov, E. P. (1978) "Seismic Behavior of Ductile Moment-Resisting Reinforced Concrete Frames", *Reinforced Concrete Structures in Seismic Zones*, ACI Publication SP-53, 2nd printing, Detroit, MI, p. 249.

8. Borg, S. F. (1983) *Earthquake Engineering*, John Wiley and Sons, New York, p. 103.

9. Mueller, P. (1981) "Behavioral Characteristics of Precast Walls", *Design of Prefabricated Concrete Buildings for Earthquake Loads*, Applied Technology Council, Berkeley, CA, April 27–28, p. 1.

10. Sharpe, R. L. (1978) "The Earthquake Problem", *Reinforced Concrete Structures in Seismic Zones*, ACI Publication SP-53, 2nd printing, Detroit, MI, pp. 28, 33.

11. ACI-ASCE Committee 326, "Shear and Diagonal Tension", Proceedings *ACI*, Vol. 59, January–February–March, 1962, 426.

12. Paulay, T. (1980) "Earthquake-Resisting Shearwalls – New Zealand Design", *ACI Journal* 77, 144–152, Figure 1.

13. Stamenkovic, H. (1978) "Suggested Revision to ACI Building Code Clauses Dealing with Shear Friction and Shear in Deep Beams and Corbels", discussion, *ACI Journal* 75, 222–224, Figure D.

14. Stamenkovic, H. (1981) "Comparison of Pullout Strength of Concrete with Compressive Strength of Cylinders and Cores, Pulse Velocity and Rebound Hammer," discussion, *ACI Journal* 78, 154, Figure B.

15. Timoshenko, S. and Young, D. (1968) *Elements of Strength of Materials*, 5th edn, D. Van Nostrand Company, New York, p. 62, Figure 3.11, p. 194.

16. Saliger, R. (1927) *Praktishe Statik*, Leipzig and Wein, Franz Deuticke, Sweiste Auflage, S. 148, Abb. 198 (p. 148, Figure 198).

17. Fintel, M. (1978) "Ductile Shear Walls in Earthquake Resistant Multistory Buildings", *Reinforced Concrete Structures in Seismic Zones*, ACI Publication SP-53, 2nd printing, Detroit, MI, p. 126.

18. Paulay, T., Priestley, M. J. N. and Synge, A. J. (1982) "Ductility in Earthquake Resisting Squat Shearwalls", *ACI Journal* 79, 259.

19. Paulay, T. (1978) "Ductility of Reinforced Concrete Shearwalls for Seismic Areas", *Reinforced Concrete Structures in Seismic Zones*, ACI Publication SP-53, 2nd printing, Detroit, MI, p. 128.

20. MacGregor, J. G. (1974) "Ductility of Structural Elements", *Handbook of Concrete Engineering*, ed Mark Fintel, Van Nostrand Reinhold Company, New York, p. 244.

21. Gallegos, H. and Rios, R. 1978 "Earthquake-Repair-Earthquake", *Reinforced Concrete Structures in Seismic Zones*, ACI Publication SP-53, Detroit, MI, 1978, p. 476, Figure 4; p. 142, Figure 7.

22. Barns, S. B. and Pinkham, C. W. (1973) "San Fernando California Earthquake of February 9, 1971", U.S. Department of Commerce, Vol. 1, Part A, pp. 227–229, 249 and 273 (Figures 35 and 36).

23. Cummings, A. E. and Hart, L. (1959) "Soil Mechanics and Foundations", *Civil Engineering Handbook*, McGraw-Hill Book Company, New York, pp. 8–10.

24. Stamenkovic, H. (1950) "Mechanism of Crackings in RC Beam and their Prevention", *Journal of Building Industry* "Nase Gradjevinarstvo" (Our Construction), Beograd, Yugoslavia, 491–495.

25. Hawkins, N. M., Mitchell, D. and Roeder, C. W. (1980) "Moment Resisting Connections for Mixed Construction", *Engineering Journal of American Institute of Steel Construction*, 7(1) 2, Figure 2.

26. Amerhein, J. and Kesler, J. J. (1993) *Masonry Design Manual*, 3rd edn, Masonry Industry Advancement Committee, Los Angeles, CA, p. 89, Figure 2.22.

27. Oesterle, R. G., Fiorato, A. E. and W. G. Corley, "Reinforcement Details for Earthquake-Resistant Structural Walls," ACI "Concrete International", pp. 55–66, Figures 12 and 23.

28. Tarics, A. G. (1984) "Conference Highlights Earthquake Effects", *Journal of Civil Engineering* ASCE No. 10, **54**.

29. Richard, R. M., Allen, C. J. and Patridge, J. E. "Beam Slot Connection Design Manual", United States Patent No. 5,680,738.

30. Stamenkovic, H. (1981) "Shear and Torsion Design of Prestressed and Not Prestressed Concrete Beams", discussion, *PCI Journal* **26**, 106–107.

31. Stamenkovic, H. (1979) "Short Term Deflection of Beams," discussion, *ACI Journal* **76**, 370–373, Figure B.

32. Stamenkovic, H. "Suggested Revision to Shear Provisions for Building Codes", discussion, *ACI Journal*, October 1978, No. 10, Proceedings V. 75, pp. 565–567.

33. Stamenkovic, H. "Diagonally Reinforced Shearwall Can Resist Much Higher Lateral Forces Than Ordinary Shearwall", Proceeding of the Eighth World Conference on Earthquake Engineering, July 21–28, 1984, San Francisco, CA, Vol. 5, pp. 589–596.

Chapter 6

Mechanism of deformation of the horizontal membrane exposed to lateral forces

6.1 A brief overview of the problem

In this chapter we describe the application of the existing theory, as well as the application of our theory in the control of deformation of a shear membrane.

6.1.1 Application of the existing theory to study deformation of a shear membrane

Extensive research of the existing structural theory reveals that the horizontal shear membrane, on which lateral forces are simultaneously imposed from two sides, should be exposed to diagonal tension only in two quadrants, while the other two quadrants are exposed to torsion. Because wind or earthquakes can come from any direction, by applying the classical theory, it appears that the horizontal membrane, in order to survive action from any direction, should be guarded against both diagonal failure and torsion. Such facts are illustrated clearly in Figure 6.1 in which it becomes evident that something must be wrong with the existing structural theory. The nature of deformation of a shear membrane and the stress conditions must be evaluated.

6.1.2 Application of our theory to study deformation of a shear membrane

As Figure 6.2 illustrates, any shear membrane exposed to a load from two sides has the natural tendency to elongate along one diagonal and to shorten along the other, as a result of pure shear action. Also, such a membrane is simultaneously exposed to flexural bending, as a result of uniform loading from two sides. Such loading from two sides could also be caused in a shear wall by the wind (or earthquake) and gravitational loads caused by the next floor.

6.2 Introduction

Because of their rigidity, intrinsic advantages over steel and other construction materials, versatility, variability of its shape, durability, inherent fire-proofing and

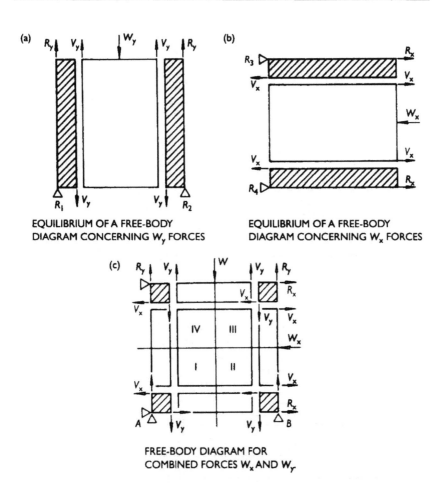

(a)

EQUILIBRIUM OF A FREE-BODY
DIAGRAM CONCERNING W_y FORCES

(b)

EQUILIBRIUM OF A FREE-BODY
DIAGRAM CONCERNING W_x FORCES

(c)

FREE-BODY DIAGRAM FOR
COMBINED FORCES W_x AND W_y

Figure 6.1 The action of lateral forces W_x and W_y on the shear membrane causes diagonal
tension in two quadrants while in the other two quadrants the lateral load
causes torsion. Shear forces shown (V_x and V_y) are *resisting* shear forces, keeping
the free bodies in equilibrium. The active shear forces (not shown) would be
oriented opposite to these.

speed of erection, concrete shear membranes (vertical or horizontal) have gained
acceptance in today's construction market.[1] Concrete buildings with concrete shear
membranes become very resistant to earthquakes: "Many concrete structures have
withstood severe earthquakes without significant damage, suggesting that there is
nothing inherent in concrete structures which make them particularly vulnerable
to earthquakes."[2]

Where concrete shear membranes are concerned (shear wall or floor), the belief
now is that "to achieve damage control, the ductile shear wall may be the most
logical solution."[3] But, because of the fallacy of the truss analogy theory,[4] the

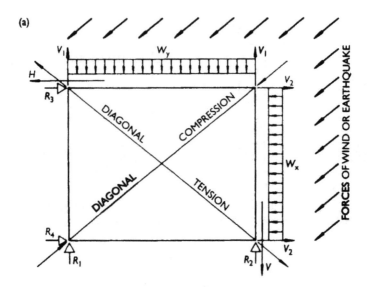

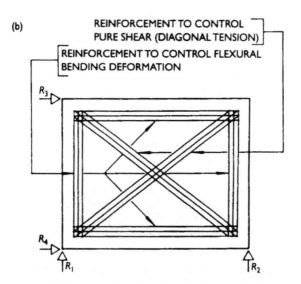

Figure 6.2 (a) Illustrates shear membrane (floor slab) exposed to the action of the wind or earthquake causing simultaneously pure diagonal tension and flexural bendings. (b) Shows that triangular reinforcement is the best protection against pure shear action (diagonal tension) and flexural bending in comparison that "it was found that horizontal wall reinforcement did not contribute to shear strength"[10,11] because such reinforcement could not control diagonal stretching (tension), diagonal compression or flexural bending.

fallacy of diagonal tension[5] and the fact that "earthquake engineering science is still an art and not a science,"[6] it appears that deformation and stress distribution in a shear membrane was not understood. Without a fundamental understanding of a given phenomenon, a permanent solution is impossible, so the main objective of this chapter is to explain the real stress distribution in a given membrane. This explanation will lead to the possible understanding of the mechanism of a shear membrane's stress condition and its deformation.

It is a clearly emphasized fact that any horizontal shear membrane is rarely exposed to wind or earthquake forces acting exclusively at right-angles to only one side, because the probability of such an occurrence is extremely slim. Actually, wind or earthquake forces attack a building at any angle (smaller or larger than 90°) during the life of the structure. By resolution of such forces into two components, a horizontal shear membrane is exposed simultaneously to two lateral forces of different magnitudes, acting separately and perpendicularly to each other and at the same time to pure diagonal tension, as is illustrated by Figure 6.2(a).

In addition, the Uniform Building Code (UBC) requires that a combination of lateral forces acting perpendicularly to each other be investigated. This is another motivation for this chapter. Section 2312(j)2C of the UBC requires that: "In computing the effect of seismic forces in combination with vertical loads, gravity load stresses, induced in a member by dead load plus design live load (excepting roof live load), shall be considered. Consideration should be given to minimum gravity loads acting in combination with lateral forces."[7] This study is justifiable from the UBC point of view alone.

6.3   Discussion

6.3.1   Newton's laws are not applicable for internal active and resisting forces for a flexural member

More than 300 years have passed since the publication of Newton's laws (1686). However, to date, to our knowledge, the laws have neither been understood nor applied correctly in the context of structural engineering theory. Newton's laws have been misunderstood, but have still been quoted in literature, for example, "equations $\sum X = 0$, $\sum Y = 0$, $\sum M = 0$, cannot be proven algebraically,"[8] though horizontal shear forces are indeed not in equilibrium as shown. Consequently, how can anyone understand diagonal "tension" failure without understanding equilibrium of a free body diagram. A free body diagram could be in equilibrium only if horizontal active shear forces (H) are in equilibrium by their corresponding resisting shear (H_r) forces. This is what Newton defined as "for every action on a body at rest, there is an equal and opposite reaction."[9] Without the proper application of active and resisting forces, it is really impossible for the equilibrium of a free body to be rationally established or understood. Because Newton's laws are not applicable where equilibrium of a free body from a flexural member is concerned, the false concept of diagonal tension has successfully survived for over a century. No one wanted to believe that any vertical shear force V must have its own resisting force V_r, in the same way as compression and tensile forces (C and T) must have their own resisting compression and tensile (C_r and T_r) forces.

Concerning the concept of "action and reaction," there indeed exists some similarity between Newton's third law and this new law, but they are fundamentally different:

1   Newton's third law is based on an external active force and an internal resisting force which tends to prevent movement of the external force.
2   The new law is based exclusively on the existence of internal active and internal resisting forces which are created in the body of a bent member.
3   While Newton's third law manifests itself throughout the universe, the new law manifests itself exclusively and only in a flexural bending and nowhere else.
4   Also, Newton's third law has nothing to do with diagonal cracking and diagonal failure in a flexural member, while the internal active forces of the new law cause such diagonal cracking and diagonal failure.
5   Newton's third law has no correlation with the new critical cross section close to supports in a bent member, while the internal active forces of the new law cause this critical cross section.
6   It is the same with diagonal cracking in a column of a rigid steel frame where Newton's third law has nothing to say, while internal active forces of the new law cause such cracking.
7   In fact, only a superficial similarity exists in the way these forces are named "active and resisting" but even here, they differ deeply because, in the new law, such forces have the additional attribute, i.e. "internal" (with precise meaning), which does not and cannot exist in Newton's law of action and reaction for the external active forces. Evidently, any similarity is indeed very superficial.

6.3.2 Concerning internal active and internal resisting forces, Newton's laws are not applicable for equilibrium of a free body from a flexural member

It is not the opinion of Sir Isaac Newton, but rather a physical law, that $\sum X = 0$, $\sum Y = 0$ and $\sum M = 0$, must be true in order for any free body to be in equilibrium. Such a fact is mathematically provable for any free body except for one, cut from a flexural member, *because internal active and internal resisting forces have not been recognized by Newton or anyone else.*

Here, it should be emphasized that translational and rotational equilibrium does exist only for vertical shear forces and horizontal compression and tensile forces because the moment of the internal forces ($C \cdot d$ or $T \cdot d$) is equal to the moment of the external force ($R \cdot x$). In other words: the couple of compression and tensile forces (C and T) is equal in magnitude, but oppositely oriented, to the couple of supporting force (R) and vertical shear force (V). (Here, x denotes the distance from the cross section to the supporting force R). Yet, because these two moments are located in the same plane and they are equal in magnitude, they are indeed in equilibrium. But, since internal horizontal shear forces H–H are in translational equilibrium, but not in rotational equilibrium, it follows that we cannot establish rotational equilibrium for this couple by Newton's first law (or his third law).

Yet, if the equations shown are provable for any other possible case, why could they not be provable for the case of flexural bending? The most reasonable answer is that Newton's first law is not applicable for equilibrium of a free body, cut from a flexural member. The reason is obvious, as shown in any textbook: internal forces C, T, H and V are not stopped by any counteracting resultant forces, so shown forces are free to move forever around our planet. Somehow, equilibrium of such a free body must be governed by some other rule of nature. And, up to now, nobody has assumed that in a flexural member, internal active and internal resisting forces are developed simultaneously, which satisfy mutual equilibrium; namely, that internal active forces (C, T, V and H) are in equilibrium by their counteracting internal resisting forces (C_r, T_r, V_r and H_r). Moreover, internal active forces tend to cause cracking and failure of the member, while internal resisting forces (as cohesion or strength of material) tend to prevent such cracking and failure.

To emphasize the importance of the above reasoning, we now investigate the stress condition of a shear membrane, as suggested by the classical theory, since stress conditions can be created simultaneously by the real action of wind and earthquake forces.

6.3.3 Mechanism of deformation of a shear membrane as the existing theory postulates it to be

To illustrate the problem of the existing theory on diagonal tension and our superficial understanding (or better, our misunderstanding) of stressed conditions in a shear membrane, we use a shear membrane, exposed simultaneously to the action of bending forces from two sides, say north and west sides.

Observe the free body diagram with the load imposed from the north side (W_y), as shown in Figure 6.1(a). The figure is divided into three free body diagrams. To achieve equilibrium of any individual free body, forces V_y must act as shown in Figure 6.1(a). Such a concept of equilibrium is recognized in any textbook; namely, that the forces V_y shown are active forces which cause diagonal tension. To show the full drawback of the existing knowledge of diagonal tension, we will not challenge, at the moment, the assumption of active or resisting forces. Rather, we will say, "OK, assume for now that such forces are indeed active forces, as has been said in any textbook or by any school in the world." With such a condition for discussion, it could be said that these three free bodies (as shown in Figure 6.1(a)) are indeed in equilibrium and that such forces V_y stipulate diagonal tension as per the existing theory.

The next step will be to observe stress conditions of the same membrane exposed to external loads from the east side (W_x) only. The three free body diagrams in Figure 6.1(b) will be in equilibrium, by the action of transversal shear forces V_x, as shown in the same figure. Such a fact is also recognized in any textbook.

The third step would be to show stress conditions of such a membrane exposed simultaneously to forces W_x and W_y, as illustrated by Figure 6.1(c). Such a condition could be illustrated by overlapping Figure 6.1(b) and Figure 6.1(a) with all their forces (V_x and V_y) thus obtaining Figure 6.1(c). Up to now, everything seems to be alright, except that now, something appears in Figure 6.1(c) that illustrates the source of confusion in understanding: alleged diagonal tension appears in quadrants I and III, while in quadrants II and IV, pure rotation (torsion) occurs without any balancing forces.

If the classical assumption were correct; namely, that the shear forces shown in Figure 6.1 (V_x and V_y) are indeed active forces, then in a shear membrane (exposed to active load from the east and north sides by forces W_x and W_y) the action of pure diagonal tension in two quadrants and permanent rotation (torsion) without any equilibrium or any diagonal tension in the other two quadrants, would appear simultaneously, as shown clearly in Figure 6.1(c).

6.3.4 Mechanism of deformation of a shear membrane as nature dictates it to be

As can be seen from Figure 6.2, any action of wind or earthquake forces simultaneously causes two different stress conditions: (1) pure shear as the main loading condition (caused by horizontal shear forces and vertical shear forces V); and (2) flexural bending (caused by uniform load W_y and W_x).

The pure shear condition causes one diagonal of the membrane to be exposed to stretching (tension) and the other to shortening (compression). In other words, such a stress condition creates two main structural elements in a shear membrane:

1   tensile tie along one diagonal, and
2   compression strut along another diagonal.

Where the above condition of a shear membrane is concerned, it becomes evident that by applying such reinforcement to prevent elongation of one diagonal and shortening of the other, deformation of such a membrane would automatically be controlled and the main problem of the safety of the membrane would be solved.

In flexural bending, the membrane will be treated as a beam on two supports and, according to the bending load, it will determine the bending reinforcement. Because such a load could occur from any angle of the structure, our rectangular membrane must be treated as a beam on two supports from all four sides and, as such, will have flexural reinforcement all around its outside edges. With such reinforcement, we could control any possible flexural bending from any side without difficulty.

The most interesting idea of this chapter is that, by introducing diagonal reinforcement to control the elongation of any diagonal of a shear membrane (against pure shear) and by introducing reinforcement at all edges of the membrane (in order to control flexural bending), we can create a triangular reinforcement to control deformation of a shear membrane by applying common sense and physical laws.

In order to increase the safety of a shear membrane, it appears that the most rational design (with the highest degree of safety) would be triangular reinforcement, because external forces would be applied to act parallel to the reinforcement (as occurs in a truss). Consequently, the most rational application of reinforcement would be achieved. This would be the safest design, because a triangle is the only rigid geometrical configuration and any diagonal tension (stretching) could be controlled by corresponding diagonal reinforcement (at the triangle's hypotenuse). This should be compared with the existing knowledge that additional horizontal wall reinforcement does not contribute to shear strength.[10,11] Any additional bars in a triangle's reinforcement will lead to additional rigidity or closer control of the elongation of any diagonal of the shear membrane. Consequently, this provides additional resistance to diagonal cracking. If we use more diagonal reinforcement in our shear membrane, then, this membrane could fail only in flexural bending with the ductility of the membrane being totally preserved. Also, the total required reinforcement (including reinforcement in both directions) could be the same or less than that required by the classical concept, but with significantly higher safety.

A shear membrane which interacts with trusses in its body will show rigidity of trusses and will be able to control much higher lateral forces due to limited elongation of diagonals and proportional load distribution between the chords. Failure of a triangularly reinforced membrane will be governed by the yielding of tension reinforcement located in the vertical struts (ties) at the edges of the membrane. Such failure is stipulated by the frame's reinforcement at the edges of the entire membrane, forcing the membrane to behave as a beam on two supports at the yielding stage.

6.4 Conclusion

It becomes apparent that flexural reinforcement (located at the edges of the shear membrane) must be present to bring flexural bending under control. By so doing, we make such a membrane triangularly reinforced, as that diagonal reinforcement will control diagonal cracking, while flexural bending would be controlled by the reinforcement located in the triangle's abscissa and ordinate.

It appears that the most rational design (with the highest degree of safety) would be triangular reinforcement, because external forces would be applied to act parallel to the reinforcement (as any truss does) and thus the most rational use of reinforcement would be achieved. It becomes evident that the entire reinforced concrete theory is now governed by a new law of engineering mechanics.

References

1. MacLeod, I. A. (1974) "Large Panel Structures", *Handbook of Concrete Engineering*, ed Mark Fintel, Von Nostrand Reinhold Company, New York, pp. 433–448.
2. Bertero, B. V. and Popov, E. P. (1978) "Seismic Behavior of Ductile Moment-Resisting Reinforced Concrete Frames", *Reinforced Concrete Structures in Seismic Zones*, ACI Publication SP-53, Detroit, MI, p. 248.
3. Fintel, M. (1978) "Ductile Shear Walls in Earthquake Resistant Multistory Buildings", *Reinforced Concrete Structures in Seismic Zones*, ACI Publication, SP-53, Detroit, MI, p. 126.
4. Stamenkovic, H. (1981) "Shear and Torsion Design of Prestressed and Not Prestressed Concrete Beams", discussion, *PCI Journal*, 106–107.
5. Stamenkovic, H. (1978) "Suggested Revision to Shear Provision for Building Codes", discussion, *ACI Journal* 565–567, Figure Ca.
6. Sharpe, R. L. (1978) "The Earthquake Problem", ACI Publication SP-53, Detroit, MI, 2nd printing, 22, 28.
7. Uniform Building Code, International Conference of Building Officials, Whittier, California, 1979, pp. 133–134.
8. McCormac, J. C. (1984) *Structural Analysis*, 4th edn, Harper and Row, New York, pp. 17, 24, Figure 2.2.
9. Sears, F. W. and Zemansky, M. W. (1953) *University Physics*, Addison-Wesley Publishing Company, Cambridge, MA, p. 32.
10. Barda, F., Hansen, J. M. and Corley, W. G. (1978) "Shear Strength of Low-Rise Walls with Boundary Elements", R.C. *Structures*, ACI publication, SP-53, pp. 150–151, 154.
11. Park, R. and Pauley, T. (1975) *Reinforced Concrete Structures*, John Wiley and Sons, New York, p. 618.

Chapter 7

Combined conclusions

It has been proved that in any flexural bending, irrespective of the type of material under the action of bending forces, two fundamentally different groups of internal forces are simultaneously created.

7.1 First group of forces as per Newton's third law

The action of any external force on a body creates its own reaction, having a tendency to prevent the penetration of an external force through the material of the body. This condition corresponds to Newton's third law where external supporting forces R_1 and R_2 and bending forces P_1 and P_2 create their internal resisting forces (or reaction) R_{1r}, R_{2r}, P_{1r} and P_{2r} as illustrated by Figures 2.6, 2.7 and 5.1 in this book. It should be emphasized here that such internal forces are pure reactions to external forces; this is known as the law of action and reaction. These are the only relevant forces at this level. It should be emphasized that this is why, until now, there has been no formulation or explanation for the coexistence of internal active and internal resisting forces in physics. Consequently, this is a new view of physics and we will deal with it in the following section.

7.2 Second group of forces as per the new law of physics

All external forces acting on a flexurally bent member (including forces from supports and external loads) form two additional kinds of internal forces: internal active forces and internal resisting (reactive or restoring) forces. As a result of their importance on the action and causes of such action in a flexurally bent member, it is necessary to emphasize that these internal forces have no correlation with the internal forces of Newton's third law, which speaks of external action and internal reaction. The only common ground is the fact that both appear simultaneously in a bent member exposed to the action of external forces.

In a bent member (except in pure bending), four different kinds of internal active forces and four different kinds of internal resisting forces can be observed:

1   Internal active forces of a (a) compression (C), (b) tension (T), (c) horizontal shear (H), and (d) vertical shear (V).
2   Internal resisting forces of (a) compression (C_r), (b) tension (T_r), (c) horizontal shear (H_r), and (d) vertical shear (V_r).

7.2.1   Internal active forces

1   Internal active compression forces are formed in a bent member and do not appear anywhere else in nature. Such forces, with their corresponding neutral plane and stress conditions, were first explained by Leonardo da Vinci in the 15th century (1462–1519). "Leonardo's test sets forth his recognition that the beam center is unstressed (the neutral zone) and that the stresses build up from there in proportion to the distance. With that realization of what is now called the concept of the neutral plane, it becomes possible to find the theoretical level of failure in all kinds of structures and machines," *Scientific American*, September 1986, 103–113.
2   Internal active tensile forces act parallel to the neutral plane of the element, simultaneously with the compression forces. They are the greatest along the edge fibers, and are zero on the neutral axis, as Leonardo da Vinci defined them in his study of the archer's bow as a combative arm. To reiterate, such compression and tensile forces are due exclusively to bending.
      We have demonstrated, and illustrated by the original sketch in Figure 3.2, the existence of internal active compression and tensile forces and their natural directions of action. This proof is based on very simple logic: In the tensile zone, for the manifestation of a crack, oppositely oriented active tensile forces must exist, acting to the left and right of the crack. This is an uncontestable fact, and so is the existence of internal active compression force in the compression zone. The existence of internal active compression and tensile forces as well as their directions of action in the bent beam are thus physically proved. It is evident that these forces are oppositely oriented to the forces that the literature defines for the equilibrium of the free body as the "active forces" of compression C, and tension T.
3   Internal active horizontal shear forces H are the forces conditioning the sliding of the laminated wood beam through the neutral plane, forming two separate members. Their existence can easily be seen by the bending of wood elements. Since such shear does not exist for pure bending, it can be assumed that these forces are due exclusively to flexural bending, as shown in Figures 2.5(a) and (b).
4   Internal active vertical shear forces V have the tendency to cut the bent beam through any transversal section. These forces are the basis for every structural analysis, their direction of action and magnitude being defined by the

following rule: the algebraic sum of all the forces, left and right from the cross section, represents the vertical active shear force for the given cross section. This is the real tendency of the support to move its portion of the beam through the vertical plane, pushing it upward in the direction of its action, while the tendency of the external load is, for the same cross section, to move its portion downward in the direction of its action.

7.2.2 Internal resisting (reactive) forces

Internal resisting (reactive) forces have the tendency to prevent deformation and allow the bent beam, when the load is removed, to return to its original shape. For that reason, these forces are also known as restoring forces. Some particular definitions of these forces are:

1   The internal resisting compression forces C_r (r, for "reaction") are due to the material strength preventing the deformation of the beam by compression. These forces resist the active compression forces C, equilibrating their action. In contemporary literature, they are used for the so-called equilibrium of the free body, but are defined as internal active forces when, in actuality, they are resisting forces.
2   Internal resisting tensile forces T_r are due to the material strength, preventing the extension of the beam. They resist the action of the internal active tensile forces, are collinear and equilibrate one another. Their existence is proved in Figure 3.2; they are forces resisting the crack's manifestation at the beginning of the bending, allowing the cracking to occur much later.
3   Internal resisting horizontal shear forces H_r prevent, for example, a laminated wooden beam from sliding along the neutral plane immediately after the bending. However, sliding occurs later when the bending increases. These forces are due to the material strength against shearing or sliding along the neutral axis, preventing the separation of the beam into two units. Forces H_r equilibrating the internal active shear forces H lying in the same plane, are collinear; equal, but oppositely oriented.
4   Internal resisting vertical shear forces V_r resist the tendency of the support to move (cut) its portion of the beam upward and the tendency of the external load to move its portion of the beam downward in the direction of its action. The forces V_r are equal to the active vertical shear forces V, lie in the same plane and are collinear but oppositely oriented.

It should be emphasized here that, in each beam as a whole, two transversal planes exist infinitely close to each other. These planes appear when a portion of the beam is separated from the whole; one plane belonging to the left portion and the other to the right. In each of these portions, two shear forces exist, one active and the other resisting; the active force being equilibrated by its own resisting shear force, located in the same plane. In other words, the left active shear force V does not equilibrate the right active shear force V, located in the plane of the other portion of the beam. We contend that this concept, as presented in the literature, is a gross error.

We feel that a part of the introduction should be repeated here; namely, that this group of forces is not covered by any of Newton's laws, nor by any other law. Thus, the idea of the existence of internal active and internal resisting forces can be presumed to be new in physics. That is why, up to the present, the equilibrium of the free body could not be proved, where in one plane, three moments (couples) exist, so that the moment of the vertical internal forces is equal to the moment of the internal compression and tensile forces, and these two moments are oppositely oriented equilibrating one another. While the moment of horizontal forces $H–H$ achieves translational equilibrium, it does not achieve rotational equilibrium. Since the internal active forces and the internal resisting forces in the bent beam cannot be understood using the existing laws of physics, it has not been possible to algebraically prove the equilibrium of the free body cut off from the bent beam. Now, by application of these forces, the equilibrium can be proven algebraically, as $C = C_r$, $T = T_r$, $V = V_r$ and $H = H_r$.

The fact that the internal active and internal resisting forces exist, allows us to establish that diagonal tension, as defined by the classical theory, does not exist. Therefore, the stresses shown on a unit element are really the resisting shear stresses, due to the bending of the beam. Further, if the unit element is cut off from the bent element, with the real stresses as conditioned by the bending, it follows that the resultant of the diagonal tensile stresses must be parallel to the diagonal cracks. However, in our theory, the real diagonal failure is clearly explained and proved. This failure is due to the internal active compression and tensile forces in combination with internal active vertical shear forces, as illustrated in Figure 3.2.

Because of the differentiation of the internal active and internal resisting forces, it becomes obvious, that the originators of the truss analogy theory have combined the internal resisting compression force (C_r) with the active external bending force, in order to prove the existence of the strut in the truss, between the support and the external load itself. On the basis of such a mistake, they have formulated the possible formation of a truss, on which they have developed the theory for the calculation of a reinforced concrete structure. This error is illustrated in Figure 2.3. Finally, in conjunction with our theory concerning the internal active and internal resisting forces, which is provable, it becomes evident that a great difference exists between a cantilever wall used against seismic forces and a real cantilever, which is the part of the beam on two supports. And, as already stated in the introduction, when the diagonal of the shear wall is elongated or shortened under the action of seismic forces, the diagonal at the real cantilever is neither elongated nor shortened. Consequently, a new theory is formulated in this book for the reinforcing of such walls against seismic forces using triangular reinforcement.

7.3   Any flexural bending stipulates two critical cross sections, one of which is a result of the new law of physics

In flexural bending, there are two totally different critical cross sections where failure (rupture) could take place. Depending on the ratio of the dimensions of the beam and the type of loading, one of them will be more critical than the other. However, ductile failure is only possible as failure due to compression and tension. The two types of cross sections are:

1   the flexural critical cross section where the bending moment is the largest and failure is expected to occur due to exhaustion of tensile and compressive stresses; and
2   the critical shear cross section where internal active compression, tensile and shear forces become equal in magnitude and failure is expected to occur under the exhaustion of punch shear stresses. For a uniform load, such critical cross section is located at a distance from the support of approximately 90 percent of the beam's depth while, for a concentrated load, it will be located at a distance approximately equal to the beam's depth from the support.

The second critical cross section above is a result of the new law of physics.

7.4   Any inverted T beam could be designed to be very safe for bridge structures

By transferring the load of the stringer to the top of the beam, using the same peripheral longitudinal reinforcement of the flange (bracket) as bearing reinforcement, and by transferring horizontal forces, imposed on the plate by the stringer, to the opposite plate of the beam (or opposite side of the column for a corbel), all problems of punch(ing) shear, vertical shear of the flange, frictional shear, would be solved and the "very complex stress distribution in the flange"[1] would be eliminated.

If residual punch shear stresses occur for some reason, or unforeseen, sudden dynamic impact at the stringer occurs, diagonally installed hanging stirrups would be the best possible protection to the developed punch shear, because they cross almost perpendicularly at the location of the possible occurrence of punching shear (at the corner of the flange and the web of the main beam).

Using our solution of loading an inverted T beam, the problem of torsion will be greatly modified by substantially decreasing the torsional eccentricity (approximately 46 percent) so the influence of the torsional effect would be much smaller. Consequently, bridge design and safety would be much higher, which is the real motivation for this discussion.

By applying a "hanger," as shown in Figure 2.16, the load from the stringers could be transferred directly to the symmetry line of the beam. Transferring the load from the stringer, located in the tensile zone of an inverted T beam, into the flexural compressed zone of the same T beam by the suggested special hangers, is probably the best possible solution to the design of bridge structures.

7.5 Contribution to safer ductile steel frame structure design: the outcome of the new law of physics

For an upward-oriented crack to appear in a column, tension must exist in the lower flange of the beam. This is due to the alteration of the negative moment acting downward (moment at the fixed end), while the positive moment between the two inflection points is acting upward. This is the result of the horizontal vibration of the frame. It is evident that these cracks are not due to the diagonal tension of Ritter–Morsch, but due to the combination of the internal active tensile forces T with the internal active axial forces V (acting as vertical shear forces V_1 and V_2), conditioning the cracking force V_n, acting normally to the given crack. By strengthening the connections of the column and the beam, catastrophic failure of frame structures, caused by the corresponding internal active forces of compression, tension and shear through the column, can be prevented for a given zone of the seismic area, thus enabling ductile failure to occur through the beam. The hinges (PI) must also be secured and moved away from the column as far as is rationally possible.

7.6 New law of physics becomes a unique ally to prestressed concrete

For the first time in history, we can truly understand why prestressed concrete is almost immune to diagonal shear cracking and failure, as well as how we can control its diagonal cracking and failure. In fact, by prestressing we really adjust the tensile force T in prestressed concrete, which means that the resultant punch shear force V_n is under full control. Yet, if we eliminate the tensile force T, we are eliminating any diagonal cracking and diagonal failure in prestressed concrete. All this is very clearly explained in Section 2.7.

7.7 Any classical shear wall could be converted into a new triangularly reinforced shear wall

Any shear wall, irrespective of its slenderness, could be designed using three different techniques. The first is a cantilever beam with a neutral plane in its body. The second technique uses a simple cantilever truss, while the third utilizes a triangularly reinforced panel as a shear wall which allows elongation and shortening of its diagonals. However, irrespective of the slenderness of the given cantilever beam, any shear wall could be designed as a triangularly reinforced shear wall, and not as a beam. Such rationale is based on the logic displayed in Figures 5.3(c) and 5.33–5.35. A vertical cantilever beam is converted into three different shear walls controlled by a triangular reinforcement which is exposed to elongation and shortening of the diagonals.

Given that safety of any shear wall against seismic forces is of utmost importance for California, as well as for the entire world, concrete panels as shear walls have been tested in a laboratory in Los Angeles, California. Results of testing show that a triangularly reinforced shear wall resisted a load which was two times higher than what the classical shear wall resisted, both having been designed for identical external loads. Results of laboratory testing of these panels are incorporated in the book.

Now mankind will be able to build nuclear power generating plants that are two times safer when compared with existing standards for shear design and they will also be less expensive.

References

1. Mirza, J. A. and Furlong, R. W. (1985) "Design of Reinforced and Prestressed Concrete Inverted T Beam for Bridges", *PCI Journal* **30**. 112–136.

Appendix

I The existence of internal active and internal resisting forces and the new law of physics

Since our new theory will fundamentally change the thinking in structural engineering, we strongly believe that the following additional comments and information from textbooks will be very useful to the experts in understanding the new law of mechanics. As soon as one accepts the existence of internal active and internal resisting forces, one automatically accepts the new theory on diagonal tension. For that reason, we will use university textbooks to prove the existence of internal active and internal resisting forces in a flexurally bent member.

I.I Proof I: the existence of internal active and internal resisting forces, presented by Timoshenko

I.I.I Internal resisting forces of compression and tension

Forces which are shown on a free body diagram (Figure 111(c))[1] are not active forces C and T, but rather resisting forces C_r and T_r, which Timoshenko[1] clearly recognized. To quote: "The resultant force system acting on the cross section is therefore that shown in Figure 111c, namely, *a resisting couple* made up of the equal and opposite forces T and C".[1] It is an undeniable fact that "resisting couple" must mean resisting forces C_r and T_r.

Further, if such forces are "resisting," they must be created as a result of the action of active forces, or be reactions to such forces. In other words, by recognizing internal resisting forces, Timoshenko also recognized the existence of internal active forces because resisting forces could not exist without the action of active forces. Moreover, resisting moment M_r must be created by resisting forces because active forces could create only an active moment!

1.1.2 Internal active and internal resisting shear forces

A VERTICAL SHEAR FORCES

Timoshenko clearly recognized the internal active and internal resisting vertical shear forces by his statement: "But this link and roller connection alone will not counteract the tendency of the left portion to move upward relative to the right portion" (p. 94 of reference 1). Here "to move upward" must mean internal active shear force V, acting upwardly (see Figure 198 of Saliger[2] and Figure 11.4 of Bassin–Brodsky–Wolkoff[3]).

Timoshenko continues, "to prevent this relative sliding, some additional device, such as the slotted bar in Figure 111(b), will be needed and the bar will exert a *downward force V* on the left portion." This "downward force" (to prevent upward movement) must mean internal resisting force which Seely–Smith defined as "resisting shear force V_r," (see Seely–Smith, Figure 128[4]).

Further clarification by Timoshenko gives additional proof of the existence of internal active and internal resisting forces:

a   For internal active vertical shear forces, he stated: "The algebraic sum of all the vertical forces to one side of the section m–n is called the SHEARING FORCE at the cross section m–n." So the existence and direction of internal active vertical shear forces is proven statically. That is a fact!

b   In his next statement, Timoshenko stated, for internal resisting vertical shear forces, "It is to be emphasized that the shearing force thus defined is *opposite to the internal shearing force but of the same magnitude* and will be designated by the common letter V."

From the above quotes it is evident that at one plane of a given cross section, two internal vertical shear forces, oppositely oriented but of equal magnitude, do exist simultaneously. It is also evident that the "algebraic sum of all the vertical forces" represents internal active vertical shear force V, while the force which creates equilibrium with such a force is its resisting vertical shear force V_r. This is also a physical fact!

B HORIZONTAL SHEAR FORCES

Internal horizontal active and resisting shear forces (stresses) were recognized by Timoshenko in his statement: "In a solid bar of depth h (Figure 159a) there will be horizontal shearing stresses along the neutral plane n–n, of such magnitude as to *prevent* this sliding of the upper portion of the bar with respect to the lower, shown in Figure 159(b)." Evidently "to prevent this sliding" must mean the resistance of materials or resisting shear forces (stresses) as they are shown in his Figure 159(b).[1]

Besides Saliger's explanation, Timoshenko's following quote is one of the best about the existing active shear stresses (forces) and their direction of action: "Observation of the clearances around a key, Figure 160b, enable one to determine the direction of sliding in the case of a built-up beam and consequently the direction of the shearing stresses over the neutral plane in the case of a solid beam" (p. 137). He shows sliding shear forces F acting in the direction opposite to resisting sliding shear stresses as they are illustrated in Figures 159(b) and 160(b). The direction of sliding shear forces F is identical to Saliger's sliding shear stresses in his Figure 198(b).[1]

C   SPECIAL NOTES

1   If one applied the above active shear stresses (Figure 160(b)) on a unit element (cube) from the same beam, then diagonal tension would appear to be parallel to diagonal cracking.

2   If one compares the direction of shear stresses on a unit element (as any textbook shows them) with Timoshenko's shear stresses as shown in his Figure 159(b), then it becomes evident that such stresses are resisting shear stresses which can never cause any stretching or cracking.

3   It is a fact that the classical concept of diagonal tension theory is based on *pure assumption*: "During the years since the early 1900s until the 1963 Code was issued, the rational philosophy was to reason that in regions where normal stress was low or could not be counted on, *a case of pure shear was assumed to exist.*"[5]

1.2   Proof II: the existence of internal active and internal resisting vertical shear forces, by Seely–Smith

Seely–Smith's definition of internal active and internal resisting vertical shear forces is so clear and evident that no one could ever deny their existence and direction of action:[4] "The symbol V will be used to denote vertical shear and, for convenience, the forces that lie on the left of the section will here be used. The vertical shear is resisted by a shearing force on the section A–A which *is called the resisting shear and is denoted by V_r.*" Further: "For the beam shown in Figure 128a the vertical shear V for section A–A is an upward force equal to 3500 lb − 2000 lb, or 1500 lb." Everything here is self-explanatory except to emphasize that for a given cross section in Figure 128, internal active shear force V is oriented upwardly while internal resisting shear force V_r is oriented downwardly, which is normal and natural.

1.3   Proof III: internal active shear stresses and their directions of action, according to Professor Saliger[4]

Probably the earliest recognition of internal active shear forces and their direction of action was by Professor Saliger in 1927: "Shear forces in a beam in relation to bending had not been taken into consideration until now. At any cross section of the beam, transverse forces are creating sliding shear stresses τ_v. Also, at the same time horizontal sliding shear stresses τ_h are created, as shown in Figure 198b."[2]

Two important facts emerge here:

1   These stresses do have identical direction as the active shear forces F of Timoshenko (horizontal shear stresses) and Seely–Smith's vertical shear forces V.

2   If one incorporates such stresses on a unit element (cube) from the same beam, diagonal tension, caused by such active shear stresses, becomes parallel to diagonal cracking.

I.4   Proof IV: internal active and internal resisting forces, by Bassin–Brodsky–Wolkoff

The existence of such forces is supported by Bassin–Brodsky–Wolkoff,[3] where they showed in their Figure 11.4 that internal active shear forces in a simply supported beam are identical to the shear forces of Timoshenko, Seely–Smith and Saliger. In Figures 11.5 and 11.6 they showed resisting shear forces or the resistance of the beam's fibers. To quote: "In addition to the bending of a beam, there is a tendency of one section of a beam to slip past the adjacent section. This *tendency is called shear and shear forces must be resisted by the fibers of the beam.*" Their Figure 11.5 shows the resisting forces as acting downwardly. Evidently the action of shear force V must be "resisted by the fibers of the beam" or by the resisting forces V_r (as Seely–Smith call them) or resisting forces V' (as Professor Kommers called them).[6]

I.5   Proof V: internal active and internal resisting vertical shear forces, by Professor Kommers[6]

A   INTERNAL ACTIVE VERTICAL SHEAR FORCES

The existence of internal active shear forces is the basic concept of any mechanics of materials text dealing with stresses. The following quote from Professor Kommers offers proof of this: "The algebraic sum of all the loads and reactions to the left or to the right of a given cross section is called *the shearing force at that section.*" So the existence of an active shearing force and its direction of action is evident beyond a shadow of a doubt: "The algebraic sum of the loads and reactions at the left of a given cross section is called *the Shearing Force.*" (See also Timoshenko, p. 95).[1]

It should be emphasized that Professor Kommers said, "... is called the shearing force at that section," which means that these shearing forces act at any cross section, as nature intended them to, as internal active forces.

It is also evident that such forces are indeed internal because they act "at a given cross section." He further stated: "It is convenient *to represent the shearing force by the symbol V'.*" There is no doubt that such a force is the real internal active vertical shear force which is identical to Saliger's shear stresses τ_v, Seely–Smith's vertical shear force V, Bassin–Brodsky–Wolkoff's shear forces in Figure 11.4 and Timoshenko's shear forces F.

B   INTERNAL RESISTING VERTICAL SHEAR FORCES

Without the existence of internal resisting forces, the equilibrium of a free body would be impossible because vertical shear force V, at a given cross section, must be balanced by some corresponding force, which Professor Kommers very clearly visualized and explained: "The shearing force must be balanced by a force produced by the internal stresses at the section, and *this is called the resisting shear*." Further, Professor Kommers refined his statement by saying, "It is convenient *to represent the shearing force by the symbol V, and resisting shear by V*."[6]

I.6   Proof VI: internal active and internal resisting forces, by Winter–Nilson

The following quoted text of Winter–Nilson[7] is enough, by itself, to prove once and for all the existence of internal active and internal resisting forces: "The role of shear stresses is easily visualized by the performance under load of the laminated beam of Figure 2.13[7]; it consists of two rectangular pieces bonded together along their contact surface. If the adhesive is strong enough, the member will deform as one single beam, as shown in Figure 2.13a." Evidently here "bonded together" must mean the resistance to sliding relative to each other, or resisting stresses to sliding.

Such an interpretation is supported by their explanation: "On the other hand, if the adhesive is weak, the two pieces will separate and slide relative to each other, as shown in Figure 2.13b." Here, "if the adhesive is weak" must mean if resisting stresses are weak then "the two pieces will separate and slide relative to each other as shown in Figure 2.13b." Here, "slide relative to each other" must mean sliding shear forces acting in an opposite direction to resisting stresses.

The direction of action of resisting shear stresses is shown very clearly in their Figure 2.13(c): "Evidently, then, when adhesive is effective, there are forces or stresses acting in it which *prevent this sliding or shearing*. These horizontal shear stresses are shown in Figure 2.13c as they act, separately, on the top and bottom pieces."[7] And, they are resisting shear stresses ("which prevent this sliding or shearing") in the same way as the resisting shear stresses of Timoshenko in his Figure 159(b),[1] Seely–Smith's shear forces in Figure 128(b),[4] Bassin–Brodsky–Wolkoff's in Figures 11.5 and 11.6[3] and Kommers' in his Figure 43.[6] The fact of the existence of internal active and resisting forces is undeniable because they are caused by nature!

I.7   Proof VII: the classical theory on diagonal tension uses resisting shear stresses instead of active shear stresses

The discussion in Section 1.6 (Proof VI) has established a very important fact: that the shear stresses shown in Figure 2.13(c),[7] are resisting shear stresses. With this knowledge, let us assume that a unit element (cube) is exposed to such resisting shear stresses (as they are shown in Figure 2.13(c),[7] then such stresses would act on a unit element as they are shown in Figure 2.14(b).[7] Since such direction of action of shear stresses serves as the international explanation of diagonal tension theory, it is necessary to emphasize that such an explanation of the diagonal tension theory is based on pure resisting shear stresses and definitely not on sliding shear stresses. This is the bottom line of this discussion: in Figure 2.14(b), resisting shear stresses are shown, not sliding shear stresses! Consequently, two facts emerge from this:

1a These stresses are stresses "which prevent this sliding or shearing" and as such, they can never cause any stretching or tension! Yet, the existing theory is trying to explain that diagonal tension is caused by these stresses, which is evidently an absurdity.

1b From another point of view, if we apply real shear stresses (caused by flexural bending) as shown in Figure 2.13(b) by Winter–Nilson or as shown by Kommers' V forces, by Seely–Smith's V forces, by Saliger's τ_v and τ_h stresses and by Timoshenko's F forces, then diagonal tension appears to be parallel to diagonal cracking. That is a fact.

2a Yet, if we assume the statement "a case of pure shear was assumed to exist"[5] to be somehow correct, then such direction of sliding shear stresses would be literally neutralized by the opposite direction of shear stresses caused by the bending phenomenon and diagonal tension becomes nonexistent.

2b As a result of the mutual cancellation, the only explanation of diagonal cracking in a concrete member is that diagonal cracking is caused by a combination of internal active compression and tensile forces (T and C) with internal active vertical shear forces (V).

I.8 Proof VIII: the existing theory for equilibrium of the free body is unprovable using the techniques of mathematical physics

Besides McCormac's statement that the basic equations $\left(\sum X = 0, \sum Y = 0\right.$ and $\left.\sum M = 0\right)$ are algebraically unprovable[8] and Sears–Zemansky's statement that a single, isolated force is therefore an impossibility,[9] the following statement by Timoshenko will provide additional proof that the existing theory for equilibrium of a free body is simply an assumption and unprovable using mathematical physics: "Two forces can be in equilibrium only in the case where they are equal in magnitude, opposite in direction and collinear in action."[10] The forces C and T, at any cross section, or forces R and V, of a free body, can never satisfy the physical requirement for equilibrium, and could never be proved algebraically to be in equilibrium.

As far as we know, Timoshenko is the only free mind who literally negated the existing explanation for the equilibrium of a free body concerning vertical shear forces, namely that the shear force V at the left cross section is in equilibrium by another vertical shear force V at the right cross section. This explanation has been applied in many engineering schools, but here we will quote only Jack C. McCormac (*Structural Analysis*, 3rd edition, p. 22): "The corresponding forces on the right free body are *by necessity* in opposite direction from those on the left. Should the left-hand side of a beam tend to move up with respect to the right-hand side, the right side *must pull down* with an equal and opposite force if equilibrium is present."[8] In other words, the left portion of a free body creates equilibrium for the right portion and vice versa.

This classical illusion, achieved by contriving to establish equilibrium of a free body, is fundamentally negated by the following statement made by Timoshenko: "But this link and roller connection alone *will not counteract the tendency of the left portion to move upward relative to the right portion* due to the unbalance of the external transverse forces. To prevent this relative sliding some additional device, such as the slotted bar in Figure 111b, will be needed and *the bar will exert a downward* force V on the left portion.*" As can be seen, equilibrium of the left portion is established by its own forces (upward and downward forces at the same cross section) so, concerning the equilibrium of a free body, the left portion has nothing to do with the right portion. Equilibrium of these two free bodies is fully and totally independent of each other! Consequently, the existing theory and the explanation for equilibrium of a free body are fundamentally false and have misled generations from successfully establishing real equilibrium of a free body.

While the existing theory for equilibrium is mathematically unprovable, the existing theory for diagonal tension is also unprovable. Our theory overcomes all such shortcomings and universally explains the causes of diagonal cracking in a flexurally bent member, in concrete or any other material.

I.9 Proof IX: modern statical proof on the existence of the free body equilibrium affirms the existence of internal active and internal resisting forces

From the generally known equilibrium conditions of the element (block) A (Figure A.1) from the bent member, applying the same horizontal shear force, the following facts can be established:

1   This force has the opposite orientation relative to the active shear force, located in the neutral plane; so that this force must be *resisting* and in no way active.
2   There exist internal horizontal *active* and horizontal *resisting* shear forces, as can be seen from the sketch.
3   Active horizontal shear forces exist in the upper as well as in the lower zone of the beam, and they condition the *couple*.
4   This *couple* is in translational equilibrium, but not in rotational equilibrium, and the equilibrium of the above *couple* is unachievable.

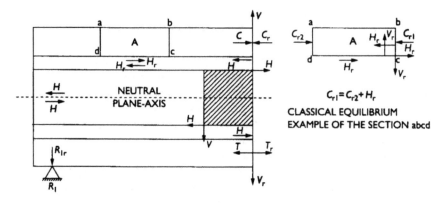

Figure A.1 Internal forces in the bent beam; for the direction of action of the horizontal shear forces see Figure 2.5(b) (Saliger).

5   When horizontal internal active and horizontal internal resisting shear forces exist, there must also exist internal active and internal resisting vertical shear forces.

6   By application of internal active and internal resisting forces, the equilibrium of the above body is achieved.

7   According to the classical theory, if diagonal tension exists, it would be parallel to the diagonal cracks, as shown in Figure A.1.

8   This one, horizontal shear force H_r from the classical theory, as shown on block A, persuasively proves the error of the whole theory of diagonal tension, in general.

9   The classical rule itself, in the proof of equilibrium of the free body (block) abcd, proves that the free body, abcd, is in translational equilibrium, but not in rotational equilibrium, since the resultant of the left forces C_{r2} and H_r, conditions the couple with the force of compression C_{r1} on the right. Or briefly, the equilibrium of the free body, cut out from the bent beam, cannot be scientifically proven, yet is *convincingly* confirmed by Figure A.1.

It is to be emphasized as well that exempting the members of Committee ACI–ASCE No. 326,[11,12] it is very difficult to meet any lecturer who will admit that Ritter and Morsch's[5] concept of diagonal tension is only a conditional hypothesis, in order to explain the diagonal failure of the bent concrete beam. This is why modern authors are beginning to use the contemporary static proof of the existence of equilibrium of the free body (Figure A.1) for the explanation of the creation of diagonal tension, as illustrated in the unit element A. And, since only a very limited number of authors are knowledgeable about the Ritter–Morsch theory which stipulates the origin of diagonal tension, supposing the existence of pure shear in a flexurally bent member, this figure (Figure A.1) seems to be an ideal example to explain such origin of diagonal tension, as illustrated by the unit element, A. To make the irony still greater, in the 1993 edition of the ACI Institute, the manual *Masonry Designer's Guide*, pp. 8–49; "Derivation of Horizontal Shear Stresses," illustrates its assertion of the element A in Figure A.1. The idea is correct. Oriented horizontal shear forces on this horizontal section actually exist but, unfortunately, it is not clear that this equilibrating force is the resisting shear force allegedly causing the cracks in the beam, and not the shear force itself. It is clear from Figure A.1, that this shear force (H_r) is oppositely oriented to the active (real) shear force (H), located in the neutral plane of the bent girder, so that the force applied on the unit element A must be the resisting shear force (H_r).

Finally, if we cannot understand a phenomenon, we must not accept and use an untruth as a truth. Nor should we utilize stereotypical terms in order to disguise our ignorance. An example of this is the statement: "The diagonal tension stresses state, before the manifestation of cracks themselves, becomes complicated, so that cracks do not follow our equations nor our foreseeings."[11,12] The truth is that we do not have the slightest idea why such a crack occurs; it is not that the phenomenon "becomes complicated".

This concludes our considerations which established the fact that resisting forces cannot condition any cracks and, yet, since diagonal cracks do appear, it is obvious that there must be other forces. It has been our endeavor to make such forces, since they are conditioned by the nature of the bent member itself, evident and understandable.

1.10 Brief comments

The evidence for the existence of internal active and internal resisting forces is so overwhelming that it simply cannot be denied. The accuracy of the statement above is also supported by the law of physics that "a single, isolated force is therefore an impossibility" (Sears–Zemansky).[9] This means that a single force C, T, V or H cannot exist without some counteracting resultants which have the same line of action. For that very reason "these equations $\left(\sum X = 0, \sum Y = 0 \text{ and } \sum M = 0 \right)$ cannot be proved algebraically" (Jack C. McCormac)[13] because no one ever fully understood and applied internal active and internal resisting forces.

The statement by Sears–Zemansky that a single, isolated vertical shear force cannot exist alone (also meaning that internal active and internal resisting forces exist) is supported by Professor Kommers: "The shearing force must be balanced by a force produced by the internal stresses at the section, and this is called the resisting shear."[6] If the resisting shear force were not produced, the vertical shear force would move in space forever.

An almost identical argument is given by Timoshenko: "To prevent this relative sliding, some additional device, such as the slotted bar in Figure 111b, will be needed and the bar will exert a downward force V on the left portion."[1]

Probably the clearest argument in favor of the Sears–Zemansky statement that a single force at a free body cannot exist (that is, that internal active and resisting forces must exist) is given by Seely–Smith: "For the beam shown in Figure 128a, the vertical shear V for section A–A is an upward force equal to 3500 lb − 2000 lb, or 1500 lb, and hence the resisting shear V_r on section A–A (Figure 128b) is a downward force of 1500 lb."[4]

If one adds Timoshenko's statement that "two forces can be in equilibrium only in the case where they are equal in magnitude, opposite in direction and collinear in action,"[10] then it becomes evident that compression force C never could be in equilibrium with tensile force T because they are not located in the same line of action. So, it is obvious that a single force (C, T, V or H) could not exist on a free body, which means:

a   that equilibrium of a free body has never been established by any author until now; and
b   that internal active and internal resisting forces do exist in any flexurally bent member.

Yet, Timoshenko's shear stresses in Figure 159(b) are oriented in one direction while Saliger's shear stresses in Figure 198(b) are oriented in the opposite direction and if they are both correct (as they are) then two oppositely oriented shear stresses do exist simultaneously in one and the same plane at the bent member. Such oppositely oriented stresses must be internal active and internal resisting shear stresses – they are real and their existence proved by experiments, and illustrated by the cited figures.

I.II   Retrospection on the psychology of man

If a person has been taught to observe the manifestation of a phenomenon in a certain way, it is very difficult to reprogram this person to observe the given phenomenon in another way. In other words, if one has to be able to accept a completely new idea, it is necessary to exclude, for a moment, all previous knowledge regarding the subject. One must set acquired knowledge aside and commence the learning process from the very beginning just as if one had never heard of the subject. Without this technique, a new idea concerning an existing phenomenon would never be accepted.

A final note: The mind of a human being is the finest mechanism which nature has ever created. Consequently, if such a mechanism is focused exclusively on one subject or problem, it can penetrate so deeply as to be able to see into a given problem, that contemporaries would not believe it possible (as in the case of Galileo and many others). In other words, the problem of diagonal cracking in a concrete member exists only because no one has ever understood the problem: "Despite all this work and all the resulting theories, no one has been able to prove a clear explanation of the failure mechanism involved" (Jack C. McCormac, *Design of*

Reinforced Concrete[13]). As soon as the problem has been understood, the problem will immediately disappear (as shown in this work).

References

1. Timoshenko, S. and McCollough, G. H. (1949) *Elements of Strength of Materials*, 3rd edn, 6th printing, D. Van Nostrand Company, New York, pp. 93–96, Figure 11; p. 137, Figure 159; pp. 94, 95; p. 148, Figure 172b; p. 72.

2. Saliger, R. *Prakticna Statika (Praktische Statik)*, Wein 1944, Yugoslav translation Nakladni zavod Hrvatske, 1949, Zagreb, Yugoslavia, p. 176, Figure 209, 210.

3. Bassin–Brodsky–Wolkoff (1969) *Statics Strength of Materials*, 2nd edn, McGraw-Hill Book Company, New York, p. 250, Figure 11.4, 11.5.

4. Seely, F. B. and Smith, J. O. (1956) *Resistance of Materials*, 4th edn, John Wiley and Sons, New York, pp. 125–128, Figure 128.

5. Wang, C. and Salmon, C. (1965) *Reinforced Concrete Design*, International Textbook Company, Scranton, Pennsylvania, p. 63.

6. Kommers, J. B. (1959) "Mechanics of Materials", *Civil Engineering Handbook*, 4th edn, Editor-in-chief Leonard Church Urquhart, McGraw-Hill Book Company, New York, p. 3–34, Figure 43.

7. Winter, G. and Nilson, A. H. (1973) *Design of Concrete Structures*, 8th edn, McGraw-Hill Book Company, New York, p. 62, Figure 2.13, 2.14.

8. McCormac, J. C. (1975) *Structural Analysis*, Harper and Row, publisher, New York, pp. 15, 22.

9. Sears, F. W. and Zemansky, M. W. (1953) *University Physics*, Addison-Wesley Publishing Company, Cambridge, MA, p. 33, paragraph 3–3.

10. Timoshenko, S. and Young, D. (1940) *Engineering Mechanics*, 2nd edn, McGraw-Hill Book Company, New York, p. 5.

11. ACI-ASCE Committee 326, "Shear and Diagonal Tension", Proceedings, *ACI*, Vol. 59, January–February–March, 1962, pp. 3, 7, 18, 21.

12. Joint ASCE-ACI Committee 426 (June 1973) "The Shear Strength of Reinforced Concrete Members", *Journal of the Structural Division*, pp. 1117, 1121.

13. McCormac, J. C. (1986) *Design of Reinforced Concrete*, 2nd edn, Harper and Row, publishers, New York, pp. 190, 191.

Index

Milton Keynes UK
Ingram Content Group UK Ltd.
UKHW040105071024
449327UK00019B/833